A Smithsonian Nature Book

Red Fox
The Catlike Canine

J. David Henry

Smithsonian Institution Press

Washington, D.C.   London

Copy Editor for this edition: Eileen D'Araujo
Production Supervisor: Kenneth Sabol

⊗ The paper used in this publication meets the minimum requirements of the American National Standard
for Permanence of Paper for Printed Library Materials
Z39.48-1984.

For permission to reproduce illustrations appearing in
this book, please correspond directly with the owners
of the works, as listed in the photo credits, p. 176. The
Smithsonian Institution Press does not retain reproduction rights for these illustrations individually, or
maintain a file of addresses for photo sources.

Cover image: Stephen J. Krasemann/DRK Photo.

The Library of Congress has previously cataloged this
book as follows:
Henry, J. David
Red fox.
(A Smithsonian nature book)
Bibliography: p. 153.
Includes index.
1. Red fox—Behavior. 2. Mammals—Behavior.
I. Title. II. Series.
QL737.C22H46   1986   599.74'442   85-43237
ISBN 1-56098-635-2

British Cataloguing-in-Publication Data is available.

02 01 00 99 98   5 4 3
Manufactured in the United States of America

Contents

Dedicated to
JANE MORAN HENRY
who will be walking with a young boy
along unexplored beaches
forever in my mind

New Preface with Updated Information on Fox Research

It has been a decade since this book was published, and recently I have received inquiries from various parts of the world asking where the book is available. Thus it is a great pleasure to see this new edition of the book published by Smithsonian Institution Press. I wish to thank Peter Cannell, Editorial Director and Science Editor of the Press, for spearheading this effort and Keith Jensen, Shelley Pruss, and an anonymous reviewer for their helpful comments.

I wrote this new preface with two goals in mind: to summarize some new research findings of fox biology from the past decade, and to share some reflections and convictions concerning field studies of free-ranging mammals.

Let's begin by looking at some of the interesting fox research carried out during the past ten years. In this brief introduction, I can only touch on a limited selection of research findings from the considerable amount available. To attain this goal, I must use a format that is somewhat different from the one found in the rest of the book. In this preface, numbers in parentheses correspond to the numbered references given at the end of this introduction. This system is the easiest way to guide the reader to the relevant research articles briefly discussed here.

When I look at the recent research on foxes, the overall theme that emerges is how fox society intersects with human society. In a broad sense, it is the diverse number of places where foxes influence humans or vice versa that has become the foci for active research (24). More specifically, during the past decade we have seen interesting research concerning:

- how foxes are related to other canids including the domestic dog;
- an assessment of the risks of rabies and new ways to prevent the spread of this disease;
- how fox social organization responds to variation in resource dispersion;
- the ecology of certain rare or endangered fox species;
- how exceedingly difficult it is to re-establish extirpated foxes in certain habitats.

Regarding terms, "canine" is a familiar word but slightly ambiguous because it can refer to pointed teeth or to members of the dog family. "Canid" is a more exact term referring to members of the dog family or their traits; I prefer to use this term throughout the book.

Evolutionary Relationships among Foxes and Related Canids

Certainly one of the great contemporary developments in biology is our ability to analyze genetic material. DNA fingerprinting techniques have made large contributions to many disciplines including biology, medical sciences, criminology, and family law. The ability to analyze an individual's genetic makeup often on a gene-by-gene basis using a drop of blood or a strand of hair has advanced our knowledge in many areas of science including canid systematics (see Chapter 2).

The degree that two species are related can often be measured by DNA Restriction Fragment Sequence Analysis; that is, by measuring the melting temperature shown by paired DNA strands. In essence, the technique works like this: Paired strands of DNA taken from the same species fit together well; these paired strands form many weak hydrogen bonds between them because of their genetic relatedness. DNA strands from distantly related animals do not fit together neatly, form fewer hydrogen bonds, and thus "melt" (separate into single strands) at a lower temperature. The melting temperature thus gives a measure of the degree of relatedness of the two animals whose DNA fragments have been experimentally paired (44).

We can now base our judgment calls about the evolutionary relationships among species not only on teeth shape, behavioral similarities, and skull and body measurements but also on the degree of relatedness of matched DNA strands. Canid systematics has been an active research area over the past two decades (5, 15, 43, 44, 45). Research has been stimulated because of dissatisfaction with the traditional view of the evolutionary tree of the dog family Canidae (40).

The order Carnivora, which includes dogs, cats, hyenas, bears, weasels, seals, mongooses, and civets, is believed to date back some 40 to 60 million years. Canidae became a recognizable group of carnivores early and is believed to have originated approximately 50 to 60 million years ago (37). New species of canids have been arising ever since. Certain genera of the foxes, such as the gray fox (*Urocyon*), have been known in the fossil record for four to six million years (32), but other fox species are of much more recent origin. For instance, the arctic fox is believed to have originated from the swift fox only 250,000 years ago, and it is probably the youngest vulpine species in existence (15, 24, 39). By comparison, even though fossil evidence of the domestic dog is rare, DNA research suggests that the domestic dog is a clear descendant of the wolf. Limited fossils and DNA research further suggest that the domestic dog as a species originated only 15,000 years before present (44).

Recent research on the DNA of Canidae (15, 44) supports that there are 34 canid species in this taxonomic family. The research suggests that Canidae should be subdivided into four groups: (a) wolf-like canids involving the domestic dog, gray wolves, coyotes, and jackals; (b) the South American canids including species such as the bush dog, crab-eating fox, and maned wolf — species of diverse morphology but quite recent ancestry; (c) the red fox–like canids of the Old and New Worlds; and (d) monotypic species such as the bat-eared fox and the raccoon dog, both of which are distinct species that have been separate from the rest of the family for a long period of time (44).

Regarding the details of fox systematics, the current view is that there are approximately 21 species of foxes from different parts of the world. They are organized into three main groups: (a) the 13 vulpine fox species of the Northern Hemisphere, (b) the seven *Dusicyon* foxes of South America, and (c) the bat-eared fox (*Otocyon megalotis*) of southern and eastern Africa. Regarding the vulpine foxes, they are traditionally grouped into four genera: *Urocyon* (the gray fox of North and Central America), *Fennecus* (the fennec of northern Africa and Arabia), *Alopex* (the arctic fox of the circumpolar regions), and *Vulpes* (red foxes, swift foxes, kit foxes, and a number of other Old World fox species). However, new evidence from DNA research suggests that *Vulpes, Fennecus,* and *Alopex* are genetically so similar that they should all be part of the same genus, *Vulpes,* but *Urocyon* merits being a distinct genus (15). This DNA research also suggests that the fennec appears to be closely related to Blanford's fox, *Vulpes cana.*

The fennec and Blanford's fox are two members of a group of foxes called desert or sand foxes. This large group of foxes is based on ecological similarities, not genetic relatedness. Desert foxes comprise nearly two-thirds of the vulpine foxes (8 out of 13 species): four from northern Africa or the Middle East—the fennec, Blanford's fox, the pale sand fox (*Vulpes pallida*), and Ruppell's fox (*Vulpes ruppelli*); the Cape fox (*Vulpes chama*) of southern Africa; the corsac fox (*Vulpes corsac*) of the steppes and deserts of central Asia; the Tibetan sand fox (*Vulpes ferrilata*) of the Himalayan foothills; and the kit fox (*Vulpes macrotis*) of northwestern Mexico and the southwestern United States. Desert foxes frequently exhibit the same cluster of adaptations that allow them to cope with the harshness of their environments. This adaptive syndrome may include a light-colored pelage for camouflage, largish ears for hunting insects and other prey as well as thermal regulation, highly concentrated urine to reduce water loss, pigmented eyes to protect against solar glare, and hair-covered feet for protection against heat and friction and to give traction in shifting sands. Recent DNA research suggests that these desert adaptations have evolved on two, perhaps three, separate occasions among this group of fox species (15).

These findings are interesting, but other taxonomic questions about foxes seem to have escaped resolution. For example, the long-term debate concerning whether the kit fox and the swift fox are subspecies or two separate species has not been resolved. Recent research remains divided; for example, DNA fragment analysis (37) supports splitting these foxes into two separate species, but morphometric and protein electrophoretic research (9) supports lumping them together as one. One begins to wonder if this debate, which spans more than 35 years, will ever be resolved.

The Campaign against Rabies

Another area of fox research that has recently received considerable attention involves the efforts to win the war against rabies (10). This endeavor has been particularly active in Britain and western Europe, but also in southern Ontario. Rabies is a virus that can infect any warm-blooded animal, but foxes, skunks, raccoons, and bats are some of the most common wildlife carriers. Rabies is usually passed on when a rabid animal bites another animal or person. The rabies virus then infects the nerves and begins to multiply, growing into the brain where the infection proliferates. Death is the usual outcome. Over the past 50 years, rabies in wildlife, principally in foxes, has been spreading from eastern Europe across a 2,000-km front into western Europe (36). There has been considerable effort to stem the tide.

Britain has been quite successful in this campaign. It has been able to maintain its disease-free status even though there have been several outbreaks of rabies, such as when servicemen returned from World War I. However, by vigilantly culling potentially infected animals, the British have always been able to eliminate the disease before it has become established in wildlife populations (21). In Britain several programs, including mandatory quarantine for all domestic pets brought into the country, are aimed at controlling the threat of rabies.

On mainland Europe several approaches against rabies have been tried with various degrees of success. Attempts to control red fox populations have only exacerbated the problem. Removing large numbers of red foxes from areas of western Europe has been found to stimulate the dispersal of red foxes from eastern Europe and thus increase the spread of the disease. Much more fruitful has been the development of vaccination baits in which pellets of fat and fish meal surround a packet containing live but weakened rabies viruses. As a fox bites and chews on the bait, live rabies viruses flood into its mouth and immunize the fox against rabies. These baits have been spread by the millions along the 2,000-km front in Europe, and organizers are hopeful that if the program is continued for the next few years, rabies could be eradicated from western Europe (2). A similar approach has been used in southern Ontario to lower the incidence of rabies and contain it to a few specified areas.

A secondary problem that arises out of the campaign against rabies involves red foxes that live in urban and suburban areas (8, 20). Britain has a particularly high urban red fox population (19), where some cities and towns support five times the number of red foxes that occur in rural parts of the country (21). Researchers are concerned that, if rabies ever became established in these urban fox populations, the potential for transferring the disease to humans would be considerable. Food resources for foxes are rich in these cities. Stephen Harris has documented that 70 percent of the diet of urban foxes in Bristol, England, originates from handouts from local residents who take great pleasure in feeding "their" foxes. Other factors make this problem even harder to manage. Experts estimate that to eradicate rabies in a fox population, over 90 percent of the foxes must be immunized. However, in urban and suburban areas, because the food resources are so plentiful, less than 40 percent of the foxes are receptive to taking baits set out for them (2, 21).

Adaptable Social Organization

Urban populations of red foxes have been the focus of considerable research that underscores the amazing adaptability of the species. For example, in pristine habitat, red foxes have usually been shown to organize themselves on nonoverlapping family territories (see Chapters 3 and 6). On each territory, a dog fox, vixen, and perhaps a "helper" female from a previous litter exist on the territory and raise the pups parented by the dog fox and vixen each year (see Chapter 3). In food-rich urban and suburban areas, social organization changes. For instance, in and around Oxford, England, red foxes live in small social groups composed of one adult male and two to five adult vixens (34). Patrick Doncaster and David Macdonald have studied red foxes living in the city of Oxford in detail. They consistently find these foxes to be living in small social groups and defending territories that vary little in size, although they change in location (6). Specifically, these urban fox territories remain relatively constant in size, and the foxes defend the boundaries, but the territories shift in location over approximately 30 to 40 hectares each year. As far as I am aware, such movable territories have never been observed among red foxes living in more natural habitat.

Extensive field research has caused David Macdonald and others (27, 31) to develop the Resource Dispersion Hypothesis. Macdonald developed this hypothesis while studying red foxes but has applied it to a number of carnivore species (35). The hypothesis is built on the fact that prey availability varies not only spatially but also temporally, not only daily but also seasonally. As a consequence, the smallest territory that will reliably support a pair of foxes, and can be defended by them, may indeed support additional foxes as well. Thus the hypothesis predicts that territory size and configuration may be determined by the dispersion of food patches whereas group size may be determined by the richness of these patches (27). It also predicts that additional foxes will be tolerated, or even encouraged, any time the costs to the dominant pair of foxes on the territory are outweighed by the benefits to them (particularly in terms of increased inclusive genetic fitness) (7).

Hersteinsson and Macdonald illustrate the Resource Dispersion Hypothesis with an example involving arctic foxes (27). A raft of eider ducks may settle in one of several coves, and a seal carcass may be washed ashore anywhere along the beach. To ensure that food is always available, one pair of arctic foxes must defend several potentially productive coves (patches). Thus the minimum territory necessary to support one pair may also support other close relatives. Applying this hypothesis to red foxes, Doncaster and Macdonald state that competitive pressure among foxes in an environment with a fine-scale patchiness of food resources (for example, in the food-rich habitat of Oxford) may favor larger territories for the resident foxes than are required by a single pair, with the result that social groups of four to five adult foxes can be supported in each territory (7, 30). On the other hand, in more homogeneous habitats, the minimum territory size that supports a pair may not support additional adult foxes. Similar patterns in social organization of red foxes have been found on the other side of the world. In southern Japan, researchers have found that nonoverlapping territories containing only a pair of foxes were the rule among red foxes in rural areas, but in urban areas adult females lived in groups of up to five individuals while male foxes maintained exclusive nonoverlapping ranges (4).

6

Little-Known Foxes

One of the most interesting developments in recent fox research is that a number of fox species have been studied in detail for the first time. Although several research programs could be highlighted (28, 29, 33, 46, 47, 48), I would like to focus on two of them, one in Israel and the other in western Canada.

Eli Geffen and David Macdonald have spearheaded a research effort to decipher the ecology and behavior of Blanford's fox, *Vulpes cana*. This is a small desert fox species, weighing approximately 1 kg in body weight. Previously it had only been known from scattered locations in Afghanistan, Pakistan, and Iran, but it recently has been discovered in Israel, and it is in the Central Judaean Reserve near Ein Gedi that Geffen and Macdonald's research team has concentrated their efforts. Among the interesting things they have discovered about this fox species are the following:

- Depending on foraging opportunities, the diet of Blanford's fox varies from being highly insectivorous to being highly dependent on fruits (13). Melons, grapes, and other pilferings from agricultural areas are frequently found in the diet of these foxes.
- It shares with a number of other desert fox species the ability to survive for long periods without drinking water. In fact, its main water source for much of the year is the water found in the plant foods that it eats (14). This restricted availability of water combined with the fox's physiology that conserves water in numerous ways allows Blanford's foxes to occupy desert areas that are far too dry for red foxes or other Old World fox species (see Chapter 2).
- Blanford's foxes show a fairly typical foxlike social organization of nonoverlapping family territories of approximately 1 to 2 km² in size (11). However, they appear to be strictly nocturnal to avoid heat stress during the daytime, avoid predation by diurnal raptors, and synchronize their activities to times of greatest prey availability (12).
- Unlike other foxes, Blanford's fox never seems to cache its surplus food (see Chapter 6 for an analysis of this behavior in the red fox). Geffen tried to stimulate food-caching behavior in these foxes by offering several of them large amounts of food. The foxes either consumed the food on the spot or carried it away and ate it, but the food was never cached (13). It is interesting that food-caching behavior also seems to be lacking in another insectivorous fox, the bat-eared fox of southern and eastern Africa (42). The absence of food-caching behavior in both of these fox species may be a consequence of feeding on small prey that are low in energetic value and that are neither easily stored nor easily recovered (13).

- As mentioned above, Blanford's fox appears to be closely related to the fennec. Both species inhabit desert environments. The fennec occupies a habitat in shifting sand dunes, whereas Blanford's fox, at least in Israel, is restricted to steep rocky slopes (14). Each species shows distinct morphological adaptations for these habitats. For example, Blanford's fox has soft, fleshy toe and heel pads that are hairless as adaptations for nonslip traction on bare rock, and the fennec has furred pads for locomotion and protection from heat and friction on dunes and loose sand (15).

In North America, the swift fox (*Vulpes velox*) of the Great Plains is a threatened fox species that has been studied quite extensively during the past decade (1, 23, 25, 26, 38). The swift fox has been extinct in Canada for over 50 years, with the last documented sighting occurring near Manyberries, Alberta, in 1938 (3). In Saskatchewan and Alberta, a large research program started in 1973 aimed at re-establishing the swift fox on the Canadian prairies (22).

The swift fox is an arid grassland fox species, about half the size of the red fox (24). Its characteristic features are its small size, long black-tipped bushy tail, soft gray to tan color with orange on the flanks and legs, and a black mark on the side of its muzzle. Swift foxes are opportunistic night, dusk, and dawn hunters and scavengers, subsisting on quite a varied diet that includes ground squirrels and other small rodents, birds and bird eggs, grasshoppers and other insects, wild fruits, and carcasses from agriculture, roadkills, and other sources. A prairie jackrabbit is about the largest prey that a swift fox will hunt (24).

Swift foxes are probably our most subterranean fox species because in their dry grassland environment they use dens not only for raising young but also for shelter from wind and storms, protection from heat and cold, reduction of moisture loss, and escape from predators (38). Swift foxes use their dens throughout the year, sleeping in or near the den for most of the day. On many days of the year the mated pair share the same den (38). Swift foxes organize themselves in nonoverlapping, monogamous family territories, with the territories being somewhat larger in areas where food is scarce (24).

In its development, the Swift Fox Recovery Program has passed through three important stages (22, 23):
- The program started releasing captive-raised swift foxes on the prairies of southeastern Alberta in 1983 using a "soft release" technique (25). In the soft release approach, researchers began by building 4 by 8 m pens on the release site and holding the captive-raised swift fox pairs for anywhere from six to nine months, feeding them daily. This soft release method was found to be very labor-intensive, and the survival of the foxes after release was disappointingly low (25).

8

- The study then moved to a "hard release" technique whereby larger numbers of captive-raised foxes were released immediately onto the Canadian prairies (22, 26). This was done because researchers decided that the habitat available at the release site might be more important than the release technique. According to Charles Mamo, one of the project's principal field workers, this habitat should show three characteristics at the time of release: there must be good escape terrain, an abundance of easily captured prey such as grasshoppers, and a low density of predators, especially coyotes. Good escape terrain simply means plenty of badger holes into which the swift foxes can escape from coyotes and golden eagles until they establish escape burrows of their own. Coyotes in particular can wreak havoc on a young swift fox population. Coyotes account for 65 percent of the swift fox mortalities in which the cause of death was determined (1, 26). This hard release technique has resulted in a greater percentage of the swift foxes surviving at least a year after release.

- The study team has recently developed an important third technique. The Swift Fox Recovery Team has been comparing the survival of swift foxes raised in captivity with that of wild foxes captured in central Wyoming and translocated to Canada (1, 23). Since 1990, the team has carried out four different releases, two in the spring and two in the fall, each release involving 10 to 20 captive-raised foxes and the same number of wild American-born foxes. The results are impressive. At the end of a 12-month period, 47 percent of the wild foxes still survived compared with only 11 percent of the captive-raised foxes (1). Among those that survived, approximately 85 percent of the wild foxes gave birth to offspring during their first year on the Canadian prairies as compared with only 25 percent of the captive-raised foxes (1). When compared with captive-raised foxes, the wild foxes survive better, reproduce more quickly, and are considerably less costly to release. The team believes that this difference is due to the fact that the wild foxes are "street smart"; that is, they have had a year or more of experience surviving in their natural habitat of Wyoming. They know how to recognize predators and avoid life-threatening situations. They also know how to capture prey and how to cope with severe weather conditions found on the prairies.

The net result of this Recovery Program is that approximately 150 to 250 swift foxes are alive on the Alberta-Saskatchewan border area, with many of these foxes formed into breeding pairs (1, 23). In moist years with abundant ground squirrels and other prey, these pairs of foxes have been observed to raise more than 50 kits per year. It is a fragile start, but a good one. We may have the beginnings of a viable population of swift foxes re-established in Canada after an absence of over 50 years.

I think there is an important take-home message that the swift fox gives us, and it is this: We have to be extremely careful with the wildlife populations that we have. The swift fox program, similar to many endangered species programs, has shown how difficult and expensive it can be to re-establish a wildlife population (22). When a species has gone extinct, it is clearly impossible to bring it back. A unique part of the Earth's biodiversity has been lost. However, the swift fox experience shows that even re-establishing a population of a species, translocating animals from one area to another, can mean decades of strenuous work and millions of dollars in costs, with the chances of success often being meager. Wildlife populations do not transplant well. They are fragile heritage resources. Their population structure and social organization have developed slowly over thousands of years, and once disrupted, these fragile entities are not quick to recover. It is far wiser to take care of the wildlife populations that are in existence than to try to restore this natural heritage once it has been devastated. This, I believe, is the most important lesson to be learned from more than 20 years of effort trying to reintroduce the swift fox to the Canadian prairies.

Protected Areas and Animal Societies

The swift fox shows that wildlife populations can be extremely difficult to re-establish. Consequently, protected areas such as national parks play a crucial role in wildlife conservation. They contribute by protecting wildlife populations and their habitat. Protected areas also provide unique opportunities to carry out research on these species. Let's use my field studies on the red fox to examine these points further.

I studied red foxes in a completely protected area, Prince Albert National Park, Saskatchewan, Canada. It is a national park where logging, mining, hunting, trapping, and aboriginal subsistence harvesting are not permitted. To my mind, protected areas such as Prince Albert National Park are uniquely valuable for at least two different reasons.

First, completely protected parks such as Prince Albert allow animals to become semihabituated and lose their fear of humans. They are places where wildlife can be observed and where we can study the intimate details of their lives. In most areas wildlife respond to humans by fleeing because their populations are hunted and trapped. In other areas wildlife is attracted to humans because there animals receive food directly or indirectly from humans. Neither of these conditions is conducive to high-quality wildlife research. To study the natural, undisturbed behavior of wildlife, you need to study animals that perceive humans neither as positive nor negative stimuli but as neutral ones (see Chapter 1). You need free-ranging animals that react to humans as unimportant parts of their environment so that the animal simply carries on with its normal natural behavior. The only locations that I know of where such a relationship develops between humans and wildlife are in fully protected national parks. This relationship can be very slow to develop, and it can be disrupted easily. It makes national parks such as Prince Albert very special places indeed.

National parks are extremely valuable for a second, and perhaps more important, reason. Experts debate how long humans have been a part of the North American ecosystems, but most researchers agree that it has been less than 30,000 years. Red foxes have existed for over two million years. They have evolved a social organization to regulate their populations and partition their resources with the end result that a sustainable population of red foxes is established. We need to understand these self-regulated mechanisms as a basis for wildlife management on all our lands, and often fully protected national parks are the only areas where fully developed, undecimated animal societies exist. To my mind, national parks such as Prince Albert are uniquely valuable. I hope that they will continue to be highly valued by all segments of our society.

Animal Watching

In closing, I would like to reflect upon the fact that *Red Fox: The Catlike Canine* was based on thousands of hours of direct observation of red foxes free-ranging in the boreal forest of northern Canada. My red fox research owes much to field ethologists. The research and writing of Niko Tinbergen (41), Jane Goodall (18), Valerius Geist (16, 17), and others emphasize methods that use the direct and patient observation of animals in their natural environment as the ultimate source of information concerning an animal and its world. This was one of the tenets of my research on the red fox. It was also based on the conviction that everything that defines a species—its particular habitat and niche, its population dynamics, its set of adaptations for coping with a changing environment, even its evolutionary relationships—is expressed through its behavior. Through carefully observing animals, we will discover insights crucial to understanding the ecology of a species; by meticulously describing its behavior, we can make discoveries concerning the ecology, behavior, and evolution of a species, discoveries that are foreign to a human perception of the world. By carrying out what Tinbergen calls "exploratory watching" (see Chapters 6 and 7), we can immerse ourselves in the *Umwelt* of the species, become partially unfettered from our anthropocentric biases, and discover new dimensions and adaptations of central importance to a species. Patiently observing and pondering the natural behaviors of a species is the sextant that allows us to navigate among these fascinating, unknown regions.

The past decade has seen important developments in a number of technical research methodologies: to mention a few, discriminant multivariate analysis techniques, computer simulation of population dynamics, and DNA restriction fragment sequence analysis. These technical advances have made important contributions to our understanding of the biology of foxes as well as other wildlife species. In fact, most of this preface describes some of the valuable contributions that they have made, and many more are to follow. Yet, in our wildlife research, I would still urge a balance between these new technologies and traditional observational techniques. Tinbergen, Goodall, Geist, and others taught us the value of observing animals carefully. I would even argue that they developed this school of animal watching into a fine science. New research methodologies have not made the observational techniques of ethologists obsolete. Taken together, they should catalyze a new synergy. Technical advances should provide us with guidance of important new things to watch for, while careful observations of the natural behavior of a species will continue to generate insights and hypotheses that require detailed technical analysis.

Nevertheless I wonder if the contemporary balance that we strike among these diverse research techniques is not somewhat out of kilter. I worry that perhaps we rush to use technical devices and analytical techniques and are too quick to overlook the value of "exploratory watching" and careful description. I worry that in this era of satellite telemetry collars, G.I.S. and G.P.S. systems, and a plethora of computer simulations of wildlife populations, some of the valuable tenets for research put forth by Niko Tinbergen and the other great animal watchers are being overlooked. How many of our researchers (young or old) spend a good portion of their field time meticulously observing individual members of their wildlife population and pondering the evolutionary meaning of the idiosyncrasies that they observe? Are we making full use of this fountainhead of potential new insights? Sometimes I feel that we get carried away by our own technology, rather than combining the best of technology with the best that natural history has to offer.

To that end, I am pleased to see this book republished. I hope

An intelligent, small predator, the red fox is well adapted to a boreal existence.

Red foxes are skillful predators. Their prey includes insects, small mammals, and occasionally birds.

A red fox scavenges along a lake shore.

Red, cross, and black—the three color phases of **Vulpes** **vulpes**. *However, foxes vary within each category; for example, many black foxes take on a silvery sheen during winter when long, silver-tipped guard hairs grow up through the black summer coat. These coat colors are genetically inherited just as eye color is in humans.*

Still cloaked in their charcoal gray natal fur, young kits explore hesitantly beyond the entrance of the den.

Between four and six weeks of age kits develop a sandy-colored coat that matches the sandy loam of the den site and helps to camouflage them during their vulnerable first month outside the den.

By six months of age, a young fox is a skillful predator.

Increased alertness is the first indication that a fox is hunting.

that it tempts wildlife researchers, particularly those who are just beginning to study a species, to consider a simpler field approach as a valuable part of their wildlife research. My advice is: go out and waste a lot of time carefully observing your animals. You may be surprised at the insights you discover. You may be surprised at the undiscovered richness of wildlife societies. And you may be intrigued by the contributions that a naturalist's approach of simply observing wildlife in its natural environment can make to the understanding, interpretation, and conservation of wildlife, the animals with whom we share this planet.

References Cited

1. Brechtel, S. H., L. N. Carbyn, D. Hjertaas, and C. Mamo. 1993. Canadian swift fox reintroduction feasibility study: 1989 to 1992. Alberta Fish and Wildlife Report. Edmonton, Alberta.
2. Brochier, B., M. P. Kieny, F. Costy, P. Coppens, B. Bauduin, J. P. Lecocq, B. Languet, G. Chappuis, P. Desmettre, K. Afiademanyo, R. Libois, and P. P. Pastoret. 1991. Large-scale eradication of rabies using recombinant vaccinia-rabies vaccine. Nature (London) 354: 520–522.
3. Carbyn, L. N. 1989. Swift foxes in Canada. Recovery 1: 8–9.
4. Cavallini, P. 1992. Ranging behavior of the red fox (*Vulpes vulpes*) in rural southern Japan. Journal of Mammalogy 73: 321–325.
5. Clutton-Brock, J., G. B. Corbet, and M. Hills. 1976. A review of the family Canidae with a classification by numerical methods. Bulletin of the British Museum of Natural History (Zoology) 29: 119–199.
6. Doncaster, C. P., and D. W. Macdonald. 1991. Drifting territoriality in the red fox, *Vulpes vulpes*. Journal of Animal Ecology 60: 423–439.
7. _____. 1992. Optimum group size for defending heterogenous distributions of resources: A model applied to red foxes, *Vulpes vulpes*, in Oxford city. Journal of Theoretical Biology 159: 189–198.
8. Doncaster, C. P., C. R. Dickman, and D. W. Macdonald. 1990. Feeding ecology of red foxes (*Vulpes vulpes*) in the city of Oxford, England. Journal of Mammalogy 71: 188–194.
9. Dragoo, J. W., J. R. Choate, T. L. Yates, and T. P. O'Farrell. 1990. Evolutionary and taxonomic relationships among North American arid land foxes. Journal of Mammalogy 71: 318–322.
10. Franklin, C. 1994. The way to outfox rabies. New Scientist 142: 16–17.
11. Geffen, E., and D. W. Macdonald. 1992. Small size and monogamy: Spatial organization of the Blanford's fox, *Vulpes cana*. Animal Behaviour 44: 1123–1130.
12. _____. 1993. Activity and movement patterns of Blanford's foxes. Journal of Mammalogy 74: 455–463.
13. Geffen, E., H. Hefner, D. W. Macdonald, and M. Ucko. 1992a. Diet and foraging behavior of the Blanford's fox, *Vulpes cana*, in Israel. Journal of Mammalogy 73: 395–402.
14. _____. 1992b. Morphological adaptations and seasonal weight changes in the Blanford's fox, *Vulpes cana*. Journal of Arid Environments 23: 287–292.
15. Geffen, E., A. Mercure, D. J. Girman, D. W. Macdonald, and R. K. Wayne. 1992. Phylogenetic relationships of the fox-like canids: Mitochondrial DNA restriction fragment, site and cytochrome *b* sequence analyses. Journal of Zoology (London) 228: 27–39.
16. Geist, V. 1971. *Mountain Sheep: A Study in Behavior and Evolution*. University of Chicago Press. Chicago.
17. _____. 1975. *Mountain Sheep and Man in the Northern Wilds*. Cornell University Press. Ithaca, New York.
18. Goodall, J. 1986. *The Chimpanzees of Gombe: Patterns of Behavior*. Harvard University Press. Cambridge, Massachusetts.
19. Harris, S. 1981. An estimation of the number of foxes (*Vulpes vulpes*) in the city of Bristol and some possible factors affecting their distribution. Journal of Applied Ecology 18: 455–465.
20. Harris, S., and J. M. V. Rayner. 1986. Urban fox (*Vulpes vulpes*) population estimates and habitat requirements in several British cities. Journal of Animal Ecology 55: 575–591.

21. Harris, S., and G. Smith. 1990. If rabies comes to Britain. New Scientist 128: 20–21.
22. Henry, J. D. 1994. *How to Spot a Fox*. Chapters Publishing. Shelburne, Vermont.
23. _____. 1995. Home again on the range. Equinox 13: 46–53.
24. _____. In press. *The Foxes of North America*. NorthWord Press, Inc. Minocqua, Wisconsin.
25. Herrero, S., C. Schroeder, and M. Scott-Brown. 1986. Are Canadian swift foxes swift enough? Biological Conservation 36: 159–167.
26. Herrero, S., C. Mamo, L. N. Carbyn, and M. Scott-Brown. 1991. Swift fox reintroduction in Canada. *In:* G. L. Holroyd, G. Burns, and H. C. Smith (eds.), *Proceedings of the Second Endangered Species and Prairie Conservation Workshop*, pp. 246–252. Provincial Museum of Alberta Natural History Occasional Paper No. 15. Edmonton, Alberta.
27. Hersteinsson, P., and D. W. Macdonald. 1982. Some comparisons between red and arctic foxes, *Vulpes vulpes* and *Alopex lagopus*, as revealed by radio tracking. Symposium of the Zoological Society of London 49: 259–289.
28. Hersteinsson, P., A. Angerbjörn, K. Frafjord, and A. Kaikusalo. 1989. The arctic fox in Fennoscandia and Iceland: Management problems. Biological Conservation 49: 67–81.
29. Hiruki, L. M., and I. Stirling. 1989. Population dynamics of the arctic fox, *Alopex lagopus*, on Banks Island, Northwest Territories. Canadian Field-Naturalist 103: 380–387.
30. Kolb, H. H., and R. Hewson. 1980. A study of fox populations in Scotland from 1971 to 1976. Journal of Applied Ecology 17: 7–19.
31. Kruuk, H. 1978. Foraging and spatial organization of the European badger, *Meles meles* L. Behavioral Ecology and Sociobiology 4: 75–89.
32. Kurtén, B., and E. Anderson. 1980. *Pleistocene Mammals of North America*. Columbia University Press. New York.
33. Lindström, E. R., H. Andrén, P. Angelstam, G. Cederlund, B. Hörnfeldt, L. Jäderberg, P. A. Lemnell, B. Martinsson, K. Sköld, and J. E. Swenson. 1994. Disease reveals the predator: Sarcoptic mange, red fox predation, and prey populations. Ecology 75: 1042–1049.
34. Macdonald, D. W. 1981. Resource dispersion and the social organisation of the red fox, *Vulpes vulpes*. *In:* J. Chapman and D. Pursley (eds.), *Proceedings of the World Furbearer Conference, 1980*, pp. 918–949. University of Maryland Press. Frostburg, Maryland.
35. _____. 1983. The ecology of carnivore social behaviour. Nature (London) 301: 379–384.
36. MacKenzie, D. 1990. How Europe is winning its war against rabies. New Scientist 126: 26–27.
37. Mercure, A., K. Ralls, K. P. Koepfli, and R. K. Wayne. 1993. Genetic subdivisions among small canids: Mitochondrial DNA differentiation of swift, kit, and arctic foxes. Evolution 47: 1313–1328.
38. Pruss, S. 1994. An observational natal den study of wild swift fox (*Vulpes velox*) on the Canadian prairie. Masters Thesis. Faculty of Environmental Design, University of Calgary. Calgary, Alberta.
39. Shield, G. F., and A. C. Wilson. 1987. Calibration of mitochondrial DNA evolution in geese. Journal of Molecular Evolution 24: 212–217.
40. Simpson, G. G. 1945. The principles of classification and a classification of mammals. Bulletin of the American Museum of Natural History 85: 1–350.
41. Tinbergen, N. 1972. *The Animal in Its World: Explorations of an Ethologist, 1932–1972*. 2 vols. Harvard University Press. Cambridge, Massachusetts.
42. Vander Wall, S. B. 1990. *Food Hoarding in Animals*. University of Chicago Press. Chicago.
43. van Gelder, R. G. 1978. A review of canid classification. American Museum Novitates No. 2646: 1–10.
44. Wayne, R. K. 1993. Molecular evolution of the dog family. Trends in Genetics 9: 218–224.
45. Wayne, R. K., R. E. Benveniste, D. N. Janczewski, and S. J. O'Brien. 1989. Molecular and biochemical evolution of the carnivores. *In:* J. L. Gittleman (ed.), *Carnivore Behavior, Ecology and Evolution*, pp. 465–494. Cornell University Press. Ithaca, New York.
46. White, P. J., and K. Ralls. 1993. Reproduction and spacing patterns of kit foxes relative to changing prey availability. Journal of Wildlife Management 57: 861–867.
47. Zoellick, B. W., and N. S. Smith. 1992. Size and spatial organization of home ranges of kit foxes in Arizona. Journal of Mammalogy 73: 83–88.
48. Zoellick, B. W., N. S. Smith, and R. S. Henry. 1989. Habitat use and movements of desert kit foxes in western Arizona. Journal of Wildlife Management 53:955–961.

Foreword

This book is natural history at its modern best because it is written with confident understanding and sympathetic feeling based on extensive and laborious observations of red foxes under natural conditions. Readers can appreciate far more fully than is usually possible what it is like to be one of these fascinating animals. Informed empathy is difficult to achieve, but it is essential for an adequate understanding of complex animals.

Scientific biologists have been inhibited throughout most of the twentieth century even from thinking about whatever feelings and thoughts may be experienced by the animals they study. Yet experienced naturalists who become familiar with the actual lives of animals find it increasingly difficult to ignore the possibility that the animals they study experience, subjectively, feelings and thoughts of some sort. When animals adjust their behavior appropriately to achieve objectives that are certainly important to them, it seems likely that they feel and think about what they are trying to do. Such thoughts or feelings need not be formulated in anything like human language, with its rule-governed grammatical syntax. But vivid imagery and basic, compelling emotion may render animal consciousness useful and significant, both for the animals and for those of us who wish to understand them.

For example, in Chapter 3 David Henry describes how red foxes select sites for digging their dens and how when one male brought back prey to feed the young he discovered that his mate and youngsters had left the burrow. It is surely a rare event in the life of a fox for the female to move the young to a different burrow while the male is hunting. But the dog fox's behavior is readily understood if we assume that he thinks in relatively simple terms such as "the cubs will eat this dead rabbit," or "my mate and cubs are gone—they must be somewhere along where I can smell their footprints." In Chapter 6 we learn how complex and effective the urine marking behavior of red foxes actually is. As they search hungrily for something edible, the complex of odors they encounter provides conflicting information. This mossy bit smells mousey, that spots looks like fur. But the foxy scent means I, or maybe another fox, passed it up as no good.

In Chapter 8 Dr. Henry describes several variations in hunting behavior that strongly suggest some simple thoughts experienced by the fox. The "nap and capture technique" is a reasonable tactic to apply after a mouse has narrowly escaped down its burrow. The simple thought that by remaining very still in a sleeping position facing the hole the fox may well have another chance to catch an edible vole seems quite plausible. And the vivid description of a fox playing with a captured shrew is difficult to interpret without assuming some thinking by the fox about what it is doing. Shrews have such an unpleasant odor that foxes eat them only when quite hungry. Playing with captured prey is a common activity of both foxes and cats. We may pity the prey animal, but to understand the real world of animal behavior we must be ready to assume that the predator enjoys this sort of play. The observation that the fox tossed a shrew back close to its burrow after carrying it back some distance to where it had been caught suggests anticipation of catching it again.

Those scientists concerned with the behavior of animals have been rigorously trained to limit themselves to observing what animals do and to trying to explain why they behave in this way. The causes considered appropriate include both genetic influences molded by natural selection over evolutionary time and learning during the animal's lifetime. But to infer that the behavior is influenced in any way by subjective mental experiences such as desires or intentions has been taboo. This viewpoint seems to be based more on intellectual aversion on the part of the scientists than on objective evidence. For in many other sorts of scientific analysis causal factors are postulated and discussed at great length even though the actual events under consideration are hopelessly unobservable. For instance, the suggestion that a certain type of animal behavior results from evolutionary selection is actually an inference that in the remote past many of the ancestors of this animal succeeded in life and raised more offspring than if they had behaved in some different fashion. But how can we ever observe such behavior of long dead ancestors? Because this is impossible we make what seem to be the most plausible inferences and then seek indirect evidence to evaluate their correctness.

The same sort of inference testing can also be performed in the case of postulated subjective feelings and thoughts. But this is such an unfamiliar procedure, due to the twentieth century taboo against scientific consideration of animal or human mental experiences, that scientists are still fumbling about trying to think how to develop appropriate procedures. Dr. Henry and other field naturalists are providing raw material that points the way, but there is a very long and prickly path to be hacked through the thorny underbrush of behavioral data and theoretical considerations before a lucid perspective can be attained. Meanwhile we can enjoy the process of trying to empathize with red foxes and other animals, recognizing candidly that we cannot yet do this very accurately. But when trying to imagine what it is like to be a red fox, and when contemplating various possible thoughts and feelings they may experience, we need not be inhibited by obsolete taboos.

DONALD R. GRIFFIN
The Rockefeller University
New York, New York

Acknowledgments

The late John Peter Clark was my first scientific mentor; he taught me early that science at its best is a simple thing. If John had lived to read this manuscript, I am sure that his comments would have made it a more lucid text.

Dr. J.M.A. Swan supervised my master's research at Harvard University, and Dr. Stephen M. Herrero supervised my doctoral research at the University of Calgary. I thank them both for their guidance and encouragement; from their example I learned that science is not only an analytical procedure but also an expedition in discovery. In addition, P.K. Anderson, Cyrille Barrette, Lu Carbyn, Dick Dekker, Valerius Geist, Ilan Golani, David Hamer, Mary Henry, Stuart Houston, Devra Kleiman, Ross Lein, Paul Leyhausen, Jon Lien, David Macdonald, David Mech, W.O. Pruitt, Jr., David Shackleton, and John Theberge read portions of my research and offered valuable suggestions. I also wish to acknowledge Marc Bekoff, Lorraine Allison, and Patricia Moehlman whose research on wild canids was the basis of my own field study of how red foxes raise their young. While I am pleased to express my indebtedness to these biologists, I point out that responsibility for the scientific accuracy of the text remains solely with the author.

My doctoral research on the red fox was made possible by generous support from the I. Z. Killam Foundation; additional financial or logistic support has been provided by the Environmental Sciences Centre (Kananaskis), the University of Calgary, Parks Canada, and the Boreal Institute of the University of Alberta. I express my sincere gratitude for their assistance.

A number of people read portions of the manuscript and through their comments greatly assisted in its development. Candace Savage was particularly helpful in this regard. I am also indebted to the late Arthur Savage and to John Anderson, Ross and Donna Barclay, Stuart Heard, Anne Landry, Rob Sanders, Paul Valliere, Michael Walsh, and Robin Wensley for their contributions. I wish to thank Merv Syroteuk as well as many others at Prince Albert National Park for their interest in my research and for their cooperation. Mary Ann Zinkan and Dennis Dyck provided photographic assistance; Mary McNeil contributed editorial acumen; Lorraine Schmidt typed the manuscript, and Sandra LeBoeuf of NORTEP assisted her in this effort.

E.F. Rivinus of the Smithsonian Institution lent encouragement and valuable editorial guidance to the project, both of which are warmly acknowledged. And finally I would like to thank my wife, Suzanne, for her careful editing of the manuscript and for teaching me to appreciate, among other things, snowshoeing, the use of language, and slightly more than half of the human race.

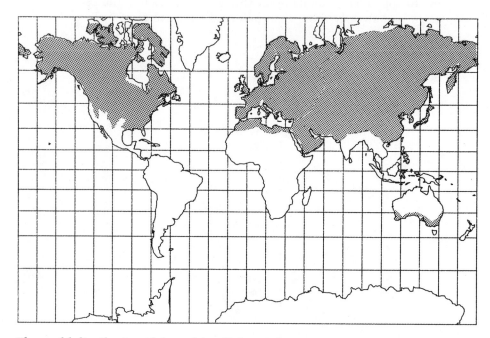

The world distribution of the red fox, **Vulpes vulpes.**

Preface: Jet-Age Foxes

Snowstorms had disrupted air travel all over the prairies, and I had to scramble to make my connecting flight. Dressed as I was for western Canada with its -30 degree weather, I was caught unprepared by the mile-long, heated corridors of the Minneapolis airport. Out of breath and perspiring heavily, I finally arrived at my assigned seat. I sank back into place and stared out the window at snow being driven past the airplane by gale force winds. After 40 minutes of waiting, the jet finally lumbered out onto the runway, taxied into position, and jarred to a halt. More delays. I did not want to think about my breakneck efforts to make the plane or about the enforced wait, so I continued to watch the snow until just past the tip of its wing, where the pavement meets the grass, a small dog emerged from the blizzard. Someone's pet, I thought quizzically. A loyal dog trying to follow its master? Not a dog at all, it was a fox. The snow let up for a moment, and I could see its luxuriant brush tipped with white and its ember-red coat flecked with snow.

The animal pranced back and forth on the edge of the pavement barking at the aircraft; the engines revved, the brakes cracked, and the jet started down the tarmac. At that moment the fox leaped forward and began to gallop down the runway. It ran not with the labored movements of a sprinting dog but as effortlessly as a leaf caught in the wind. The fox raced away from the plane, tail floating straight behind it, neck and head held proudly up-ward. It seemed almost airborne, its black, teardrop paws rico-cheting off the white, icy earth. The plane gained momentum; the snow was being driven straight across the ground now, and yet the animal kept it up—running faster and faster, cutting the storm like a wing tip.

Slowly the jet gained on the fox and gradually swept by it, and the fox, still sprinting, dropped behind and out of my view. I sat back as the jet continued to gain speed and finally lifted off. Once in the air, the plane banked to the right, and I looked out the window and down at the ground. Far below me I could barely make out a small, red, earthbound creature who was making its way back to the head of the runway where another jet waited. As the plane entered the clouds, I leaned back and thought about why it was that dogs chase cars while red foxes chase jet airplanes.

This encounter at the Minneapolis airport was one of the first times I had ever seen a red fox. Since that experience I have been repeatedly impressed by this audacious and adaptable creature. It has been using man's inventions as its playthings for thousands of years—from the vineyards of ancient Greece to the airports of the twentieth century. Man's greatest achievement, his technology, has at least proven good entertainment for red foxes.

*   *   *

Several years later, with a great deal of effort and no small measure of serendipity, I found myself traveling north to conduct a field study on red foxes. The locale for my study was Saskatchewan's Prince Albert National Park, a 1,500 square mile wildlife sanctuary. This park preserves a boreal landscape: sphagnum muskegs filled with orchids and insect-eating pitcher plants; sandy eskers, snaking out across wet lowlands; and small kames decorated with translucent birch forests. This area is also a land of lakes; one third of the park is open water made up mostly of large oval lakes interconnected by meandering creeks and rivers. Along these waterways much of the human history of the area has ebbed and flowed. This chronicle begins in the headwaters of the rich Cree and Chipewyan cultures, flows through the halcyon days of the voyageurs, and then divides and braids through the turbulent fur trading wars of the Hudson's Bay Company and the independent Metis traders.

The forest that grows throughout northern Saskatchewan is part of the taiga—the vast conifer forest that stretches across the continent from Newfoundland to Alaska. When I went up in a plane to look at this boreal woods, I had difficulty finding it because when I looked down I saw hundreds of different forests— each stand distinct. The great northern woods are really a mosaic, and inextricably woven into this pattern are the gentle, undulating roll of the land, the ravaging and rejuvenating effects of fire, and the quiltlike pattern of different mineral soils laid down by retreating glaciers.

This forest-covered plain where I had come to study foxes is not a dramatic land like the mountains to the west or the Precambrian Shield to the north; rather it has a subdued and yielding temperament. Even the processes that formed this land are not harsh and dramatic but are for the most part quiet, slow, and fluid. It is a land sculptured by water in all its various states—the grinding force of glacial ice, the slow intrusion and sedimentation of ancient inland seas, the sorting action of frost that forms and thaws with each passing winter. When a person hikes through this land, he experiences its kaleidoscope of moods from the verdant, light-filled aspen forests to the dank, quiet, threatening spruce woods. Ultimately one feels the need to guard against becoming lost in this boreal labyrinth. It is this juxtaposition of vastness and intricacy that gives the land its character.

The red foxes found within Prince Albert National Park have not been trapped or hunted for more than 50 years and as a result have lost their shyness of humans. At first I found it unnerving to be confronted by a fox that had little fear of man. I remember well one of my first encounters. I was hiking one evening along the shores of Waskesiu Lake. It was after a heavy rain, and the forest was shrouded in mist. I rose over a small hill but stopped abruptly when I saw a red fox on the trail heading my way. The fox was looking off to one side as if it were searching for something in the tall sedges that grew adjacent to the trail. I stood perfectly still, not wanting to scare this slender little predator. The fox most certainly saw me; several times it turned and casually looked at me as it trotted along. At one point the fox tensed and stood erect, its ears alert while it searched the sedges intently. I felt it was trying to locate prey among the shoots. The fox seemed so absorbed in what it was doing, so intent in search- ing among those long, damp stems. And yet in the next moment the fox had given it up completely; it began walking down the trail again, searching the other side of the path for some treasure only it seemed to know was hidden there. The fox trotted by me on the narrow trail, its back passing close to my knees. I re- mained frozen in place and watched as it trotted over the hill and out of sight. Instantly I felt I wanted to go with it, and I began to jog quietly down the path. Apparently what had transpired up to this point was well within this fox's ken, but a human who wanted to follow foxes was something new. It rocketed away from me down the trail in a breezy, light-footed gallop. I stood and watched it flee, remembering that I had seen that airy gait once before. I watched as the fox bounded off the trail and through the dense, moist woods hardly making a sound.

I had come to the park that summer to work on a black bear management project, yet something about those foxes captivated me. Whenever I could spare a moment from the bear research, I found myself observing foxes. I watched them at dawn and dusk as they hunted along the roadsides of the park. Sometimes I caught glimpses of them as they made their way through the purple-cushioned muskegs or disappeared into moss-carpeted spruce forests.

I remained behind after the bear study was finished for the season and continued to observe the foxes during the fall and early winter. After weeks of patient work, I found that certain of these foxes slowly habituated to me and allowed me to follow them through the forest—at first for only short periods but later for minutes and then hours. I found that with the right combination of stamina and patience I could follow and observe these flame-colored predators as they hunted and scavenged for a living on their boreal territories.

It was sometime during that fall that my hobby of studying foxes became a preoccupation. I was watching the foxes hunt, scavenge, and store their surplus food. I wanted to understand these behaviors in detail. Fortunately over the course of the next year my avocation of watching foxes turned into a Ph.D. research program—a project aimed at describing and analyzing the major adaptive strategies of red foxes. This research has continued since that time, occupying most of the last 14 years of my life.

1   *In the Country of the Fox*

And if our field studies have convinced me of one
thing, it is of the fact that the imagination of even
the best field biologist falls far short of the reality. . .

Niko Tinbergen
(1966)

I have watched more than a dozen generations of foxes be born in
the boreal woods. I have followed these elusive, high-strung creatures through smoky-green spruce forests and trackless muskegs.
In the first light of a winter's dawn, I have watched a fox scavenge a few drops of blood where a lynx had killed a snowshoe
hare. I have witnessed a fox stalk catlike across the forest floor
toward an unsuspecting squirrel. I have followed a fox as it
tracked a wolf who had futilely chased a frightened deer. I have
seen some of my foxes prosper and others die and have pondered
the meaning of predation in a land where the numbers are never
right, where there are always too many predators or too many
prey.

By observing wild foxes in their own land and on their own
time, I have begun to understand how the fox and its world fit
together. I have discovered some of the strange rituals and tradi-
tions that foxes observe and have slowly deciphered the rules that
foxes live by. By entering the realm of the red fox, I have gained
an expanded awareness of what it means to be alive—of what it
means to be an animal struggling to survive in one small corner
of the universe.

Studying wildlife is like making a journey to a foreign land. An
anthropologist who lives with the aborigines of Malaysia returns
with descriptions of customs and ethical values that are foreign to
the West. He or she describes religious beliefs and rituals that we
have never heard of and may be able to explain why these attitudes and customs make good sense in the tropical rain forest of
Central Borneo. By creating an appreciation of lifestyles different
from our own, an anthropologist can expand our awareness of
what it means to be human.

A wildlife biologist makes much the same kind of journey, for animals, just like man, have intricately structured societies and customs. It does not matter that the laws of red fox society are not written down in an official document; they are biologically "written down" in the brains and bodies of every red fox. Young fox cubs, partially through genetic inheritance and partially through learning, acquire these rules from their parents. The laws of the society are evident in the behavior of every red fox. It is an old and well-established culture that has been refined during the 30 million years that red foxes have inhabited the earth.

To enter the fox's world is not easy. To gain access a naturalist must at least partially leave the human world behind. In many ways a fox is the antithesis of man because usually it is a solitary creature. Foxes normally travel alone; this is how they best cope with the world. Highly social man finds this lifestyle difficult to understand let alone appreciate. Yet to understand red fox culture as antithetical to human culture is the challenge of voyaging into the country of the fox.

This is a journey into an exotic culture, and yet, unlike the anthropologist, one does not have to go halfway around the globe to enter the world of the fox. Vulpine society may exist in the nearest woodlot. As a "civilization," the red fox has been very successful. It exists in virtually every part of the Northern Hemisphere. Despite bounties, poison campaigns, and wanton destruction of dens, red fox society has flourished. Its survival is an illustration of the intelligence and ingenuity that the fox has shown in the face of persecution by man. Despite man's opposition, red fox society has remained a viable subculture that continues to thrive in many of our rural areas and even on the periphery of our suburbs. It is an intriguing foreign culture, close at hand, waiting to be explored.

2   *The Masterful Fox*

One begins to understand the red fox when one becomes fascinated by its beauty because it is the grace and elegance of the creature that have piloted its evolution. Aesthetics and adaptation are one in the red fox—its most exquisite features are also some of its most important tools for survival. For instance, the long tapering limbs, gazellelike body and yellow serpentine eyes of the fox are stunning characteristics, but they are also physical attributes that help make the animal a lethal hunter. Nor is it always easy to understand the beauty of the fox. Consider its improbable pumpkin-colored coat, black velvety ears, or magnificent brush of a tail. Why does the red fox flaunt such conspicuous features when most predators prudently emphasize camouflage?

The fox's external traits, however, are only the veneer of its beauty; its elegance is rooted in deeper levels of its life. One summer evening I watched a fox hunt mice in a meadow rich with wildflowers. The fox trotted and pounced about the blossom-laden field, tracing arcs through the sunlit air. It was a performance I shall never forget, one that convinced me that the red fox is the prima donna of wild canids—a lithe creature, yet one that moves with predacious accuracy. Anyone who has startled a napping fox and watched it flee, floating over windfallen trees and dense underbrush as if it were half bird, has begun to experience the beauty of the fox.

There is nothing heavy or lazy in a fox's movements; the fox is a wiry, highstrung creature—an animal that seems to live intensely if not for long. A fat and indolent fox seems to me a contradiction of terms—an impossible creature that could not occur or at least could not survive in the wilderness for long. A red fox will race through dense brush at speeds of 30 miles per hour, a feat very few of its prey and even fewer of its predators will attempt. It is as if the underbrush has acted as a pumice stone during the fox's evolution, shaping and polishing the creature for 30 million years until it has become one of the most agile of canids.

Yet the artistry of the red fox extends beyond its striking appearance and graceful movements. Its beauty extends to a subtler plane—to the way the various pieces of its life fit together. At this recondite level, the fox's life resembles an oriental wooden puzzle, one that when viewed initially seems to be an impenetrable block of solid wood. Taken apart it becomes a pile of disjointed carvings. Only by arduous study can one learn to assemble the parts and gain an appreciation of the intricacies and wholeness of its design.

The fox is like such a puzzle because the species is a faunal riddle that has been evolving in complexity since before man began to walk the earth. The most disjointed elements of the fox's life upon closer examination are adapted to one another with an intricate kind of animal logic; they are shaped to complement one another, often in ingenious and unexpected ways. But most of all the various facets of a fox's life entwine with its habitat, creating a life strategy that has supported the species since ancestral red foxes first appeared during early Miocene times. It is at this level that the full beauty of the red fox is expressed.

It is not easy to decipher an animal culture that has taken so long to evolve. One must use imagination to expand the human ken to understand the very different lifestyle of the fox; yet one cannot allow imagination to become excessive. An explorer of animal societies is ultimately a biologist, and as a scientist he or she must be aware that his or her ideas and conjectures are but theoretical notions, hypotheses to be tested. For example, an idea may occur to me concerning why it was adaptive for the fox to evolve as a solitary predator. Yet I must ask: Is this "fox" or isn't it? How do I test my hypothesis? How do I let the fox show me whether my conjecture is indeed a facet of its life? Ultimately the challenge is to design valid field experiments that will either entice foxes to reveal true aspects of their lives or, alternately, show the scientist that he or she is getting carried away by imagination. It is with an ever-shifting blend of imagination and skepticism that one sets off to understand the manner in which the fox's life fits together.

The Fox's Diet

The foundation for our understanding of the fox should be solid and basic; from there we can work outward. Thus we shall begin with what the fox eats. The fox's diet will form the center piece of our puzzle.

Hunters believe that foxes live mainly on grouse, quail, pheasants, and perhaps an occasional young deer. Chicken farmers believe foxes eat mainly chickens, and sheep ranchers contend

A dog fox transporting a freshly killed red squirrel back to the den.

that they eat lambs. Who is right? Accusations such as these have stimulated research by wildlife biologists into the food habits of the red fox; in fact it is the most thoroughly studied aspect of fox biology.

In surveying the literature, I found a mere 158 different studies of the red fox's food habits carried out in many different parts of the world. When these studies are taken collectively, they document two facts about the vulpine palate. First, the fox's diet is best described as catholic—the red fox eats a wide variety of food, and its diet often changes from season to season. Second, the fox is an opportunistic feeder who will sample any acceptable food often in proportion to its availability. A study done in Missouri is typical. It found that foxes had eaten 34 different mammals, 14 species of birds, 15 families of insects, and 21 species of plants. However, this same study also found that most of these animal and plant species contributed less than 1 percent by volume to the fox's diet. In this study three items—the cottontail rabbit, the meadow vole, and insects—comprised more than 60 percent of the fox's nourishment. Thus foxes are catholic in the range of what they will eat and definitely take advantage of seasonally abundant food. However, over the course of a year foxes seem to rely on a limited number of plant and animal species as their real "staples." These usually include small rodents, rabbits, insects, wild fruits, and berries. The upper size limit of prey is usually an animal no bigger than a jackrabbit, while abundance and ease of capture seem to govern whether small food items are eaten. For example, foxes eat great quantities of beetles, grasshoppers, and crickets when they are easily obtainable, and in some areas during summer and fall foxes may exist for long periods of time eating mainly berries and fruits.

This body of research shows that red foxes usually procure their food in one of two ways. The fox is an efficient scavenger but also a skillful hunter. A fox will eat or store whatever acceptable food is readily available whether it is windfallen fruit, bird eggs from an accessible breeding colony, winter-killed fish, or carrion from a road-killed deer. But the fox is also an efficient hunter specializing in capturing insects, small rodents and insectivores, rabbits, and hare.

What about the sportsman's contention that foxes hunt game birds? My observations are typical of what many biologists have observed. Foxes certainly hunt birds but not very effectively. It is rare to find more than 20 percent of the fox's diet made up of wild birds. These avian prey are usually the larger ground-nesting birds (ducks, grouse, and gulls) that the fox has caught during the vulnerable spring nesting period. But during the remainder of the year birds make up only a small portion of the fox's diet. Birds' keen eyesight, split-second reactions, and ability to fly up out of reach usually thwart even the best efforts of a fox on the hunt.

Scavenging and hunting present the fox with one problem—unreliability of food supply; that is, they are confronted periodically with times of abundance or shortage. However, they have evolved a means of partially coping with this uneven supply of food by developing an effective storage system. Whenever a fox is in possession of surplus rations (food not needed to satisfy its immediate appetite or to feed its young), the fox carefully hides these provisions away for future use. For each item the fox digs a shallow hole, pushes the food in, buries it, and then disguises the spot with leaves and twigs. Later, if the cache has not been robbed by another animal, the fox will often relocate it and make use of the food. We will study many fascinating aspects about the fox's caching behavior in a later chapter. The only point I wish to make now is that the fox has evolved a storage system to even out the distribution of its food resources.

Scavenging, hunting, caching—the three prongs of the fox's livelihood—combine with its variable diet to make this animal a most adaptable creature. Because of its flexible feeding strategy, the red fox is able to occupy a range of habitats varying from the tundra in the far North to the chaparral thorn forests in Iran. The same species of fox, *Vulpes vulpes*, that catches lemmings in the taiga forest of Canada and Siberia also harvests locusts on the edge of the Sahara Desert. In fact, the red fox has been so successful that it can be found on four of the six continents—North America, Eurasia, the northern portion of Africa, and Australia, where it was introduced during the mid-1800s and is prospering. Consequently the red fox is one of the most abundant and widely distributed wild canid species living in the world today.

The red fox is not the only kind of fox; there are in fact 14 different species of foxes occupying many corners of the world. A few of these are the arctic fox *(Alopex lagopus)* of the tundra, the kit fox *(Vulpes macrotis)* of the American Southwest, the gray fox *(Urocyon cinereoargenteus)* of the eastern deciduous forest and Central America, and the Tibetan sand fox *(Vulpes ferrilata)* of the Himalayan foothills. Because of its worldwide distribution, however, the red fox is the variety most people think of when the word "fox" is mentioned. Its wide distribution is part of the reason that Linnaeus in 1758 gave this species the scientific name *Vulpes vulpes*. With a small amount of poetic license, this Latin name can be translated "the fox's fox"—an appropriate name for this highly successful wild canid.

These main points, mentioned above, can be summarized as follows.

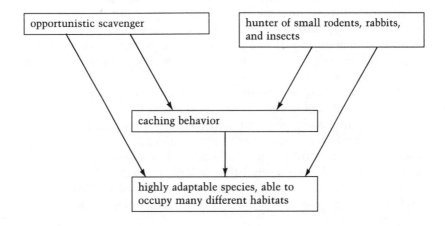

31

What do all these boxes and arrows mean? In the diagram an arrow indicates that there is an evolutionary relationship between the two boxes; specifically the factor at the base of the arrow has caused or has led to the evolution of the factor at the head of the arrow. The fox starts as both an opportunistic scavenger as well as an efficient predator of small rodents, rabbits, and insects. These are food resources that are periodically abundant and periodically scarce. Thus these two factors have led the red fox to evolve a third factor, an efficient caching system to even out the distribution of its food supplies. All three factors have caused the fox to evolve into a highly adaptable species able to occupy many different habitats.

Predator of Small Prey

Now let us consider one of the keystones of the vulpine existence: the red fox as a skillful predator of small prey. Let us reflect for a moment upon what other pieces of the fox's biology can be assembled around this important part of its life. Consider its solitary nature. I believe the fox has evolved its solitary lifestyle as a result of several factors, the most important one being the type of prey it hunts. To understand this evolution, let me briefly compare foxes with wolves.

Wolves are the largest living canids, weighing up to 175 pounds. But they often hunt prey considerably larger than themselves such as moose, elk, bison, and deer. A bull moose may weigh 1,100 pounds or more, and a mature male bison up to 2,000 pounds. These large-bodied prey are not easily killed. Many of the wolves in a pack must cooperate in order to dispatch one of these large animals, but when they are successful there is enough meat that every member of the pack gets a meal. Consequently wolves have evolved to become social predators; they hunt in packs. Wolves are very different predators as compared to red foxes. Wolves cooperate to kill and they kill infrequently—but when successful the whole pack banquets. The wolf's life then is one of feast or famine.

The red fox is not a miniature wolf. It is a different type of predator—a solitary hunter. Foxes hunt small prey, animals that a single fox can readily dispatch. Killing is the easy part for the fox; catching these alert, little prey is the challenge. Furthermore, even when successful the fox ends up with a mouse, not a moose. Because it captures an animal that is large enough to feed just one, the fox hunts alone. Try to imagine a pack of six foxes dividing up a mouse and you begin to understand why the fox is a solitary predator.

This is one reason, but a second pressure exists that has made the fox evolve into a solitary hunter. This constraint is related to but different from the one described above. Foxes' prey are high-strung and quick in their reactions. To capture these animals, the fox must stalk them quietly—remaining undetected for as long as possible—and use a surprise attack. Given a moment's warning, the prey can escape down a burrow or fly up out of the fox's reach. Thus the fox must use stealth, and to hunt by stealth the fox must hunt alone.

Pack hunters like the wolf do not often hunt by stealth; there are too many of them. A pack cannot stalk silently toward a prey and remain undetected, at least not for long. There are too many feet moving through the underbrush, too great a chance that a noise will be made thus alerting the quarry. Instead wolves usually approach prey openly, make the animal run to test its health and vigor, pursue it over a long distance, and finally pull it down. On the other hand, predators like the fox that stalk catlike toward prey and capture by surprise almost always show a solitary life-style. It seems to be a prerequisite for hunting by stealth.

Now consider two additional points related to the fox's style of predation: the fox's activity patterns and some of the properties of its sense of hearing. We will examine each point separately.

The red fox usually has been found to be crepuscular, that is, active at night but most active around dawn and dusk. Why? The reason seems to be that dusk and dawn are the times when the fox's prey are most active. Thus the fox's activity patterns mirror those of its prey. In addition, during winter mice and voles become more active during daylight hours because they live under the snow in a twilight world, one that is continuously cut off from the sun. From what has been said, one might expect that foxes should become more diurnal during winter. This in fact has been found to be the case.

What about the fox's hearing? How are its prey mirrored in the ears of a fox? A common pattern among mammals (including man) is that each species's sense of hearing is highly sensitive to the frequencies of its offspring's distress calls; but the red fox appears to be an exception to this rule. Experiments on captive foxes at Michigan State University have shown that a fox's hearing is most sensitive to lower noises (sounds ranging around 3.5 kilohertz). These frequencies correspond to the rustling and gnawing sounds that small animals make as they move through vegetation or feed on seeds, buds, and twigs. Experiments by the Scandinavian biologist Henrik Österholm have shown that red foxes can locate these sounds to within inches of their true location. I have

On occasion the prey comes to the predator, as in this case when a red squirrel in a nearby tree disturbs a napping fox.

watched foxes capture small rodents moving about under two feet of fluffy snow. To capture these quarry, the fox usually sits, appears to listen intently, and then dives into the snow, digs frantically, and either pins the prey to the ground with its paws or bites a hold of it. As a result of their finely tuned ears, foxes appear to be able to capture animals they have never seen but only heard moving under snow or through dense undergrowth.

In Search of Habitat

One more facet of the fox's life is central to its hunting behavior: what constitutes prime habitat for the species. We can most easily understand this aspect of fox biology by studying how these animals travel across the land.

Populations of red foxes do best, that is, they become most abundant, in country that is varied—land that is made up of a patchwork of woodlots, open meadows, dense brushlands, pastures, and small wetlands. The more diverse an area, the more red foxes seem to thrive in it. Ernest Thompson Seton knew this about the species; he described red foxes as "animals of half open country."

Diversified country then is prime habitat for the red fox. Why? Seton observed that a varied countryside lends itself to the varied diet of the fox. This, however, is only part of the reason. Biologists who have radio-tracked foxes for long periods have found that they spend a great portion of their time traveling along an edge of countryside where two habitats meet; for example where woods and meadows join. The fox's preference for edge environments can be understood by looking at the links among vegetation, the fox's prey, and the fox's hunting behavior. Often vegetation is more dense and diverse in the transition zone between two habitats because plants from both sides frequently grow there. As a result of this lush vegetation, small rodents, rabbits, birds, and insects often find better food, shelter, and nesting sites in these transition zones and become more abundant. Quite understandably foxes tend to hunt where their prey is abundant—consequently foxes are often observed to be "predators of the edge."

As a result of the fox's style of predation, we can add the following components to our diagram (again each arrow means that the factor that comes before the arrow has caused or has led to the evolution of the factor that follows it):

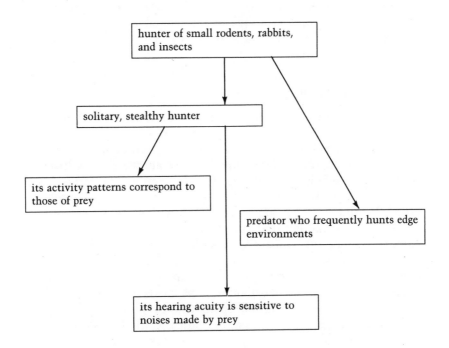

Now let us consider how much land it takes to support a fox family. The answer is, if conditions are favorable, surprisingly little. A British biologist has documented that one family of red foxes living in the rich, varied countryside near Oxford supported itself for several years while living entirely within 25 acres of land. A home range this size is quite restricted; however not all foxes attain this level of "civilization." It is more typical for a fox family to use two or three square miles for its living space. But even this amount of country is a far cry from the 100 square miles that a wolf pack normally needs for its territory. A fox family can exist on a few square miles because of its flexible feeding habits and because it can live on small prey that are often diverse and abundant compared to the density of large prey.

We are now ready to look at the fox's sense of territory. Because a family of foxes uses only a few square miles of land, it is possible for the adults to defend this area for their exclusive use. Field biologists have often found that each fox family exists on a well-defined, nonoverlapping territory that the adults, mainly the males, defend against other foxes. Within this defended area, the adults find enough food to support themselves and their growing young. The boundaries of each family territory are often well known and honored by neighboring foxes. They are further identified by the canid predilection for establishing scent posts throughout the area. Intruding foxes, such as adolescents leaving their parents' territory, are often viciously attacked and driven out of the area by one of the territorial owners.

Rules of Parenting

Let us now examine how red foxes raise their young and how their parenting style suits their lifestyle. As I have mentioned before, adult foxes support themselves and their offspring by hunting, scavenging, and caching. Because an average fox litter contains five whelps, it usually takes the combined efforts of both the male and female fox to provide enough food for the growing pups. Hunting and scavenging are hardly sedentary activities; they are demanding enough that young foxes cannot travel with their parents. Instead the kits are kept within or immediately around a den, and they rely upon this lair for their safety until they are about three-and-a-half months old. During this time the parent foxes carry food to the earthbound pups to keep them well fed.

Another consequence of both parent foxes hunting and scavenging to provide adequate nourishment for the whelps is that a strong pair bond has evolved so that both foxes of the breeding pair stay together and cooperate in rearing the young. Monogamous pair bonds have been observed in a number of animal species and apparently evolve for a variety of reasons. They may evolve as a mechanism by which the breeding pair cooperates to defend a scarce and valuable resource such as a nesting site or territory with rich food resources. Or the environment may be so difficult that it may take two adults to cope with it. For example, a solitary female may not be able to rear a litter without aid from her mate. The monogamous pair bond may facilitate early breeding that under certain ecological conditions leads to larger litters or multiple broods of offspring being raised during one season. Or the species may simply exist at a very low density, with males and females being so spaced that only a single member of the opposite sex is available for mating. Devra Kleiman and E. O. Wilson offer interesting discussions concerning the ecological conditions under which monogamy in a species is favored to develop. In the red fox, a monogamous pair bond is the usual but not exclusive observed condition. Based on the available evidence, it is not yet clear whether the pair bond among adult foxes is usually for life or for just the duration of one breeding season. Whatever the case the male fox often assists in rearing the whelps not only by defending the home range against other foxes but also by bringing food to the vixen while she is nursing and then later to the pups as they grow and develop.

An interesting offshoot of the fox's monogamous relationship is the lack of sexual differentiation among foxes. Male and female red foxes show the same markings, and male foxes are only slightly larger than females and have only slightly broader heads. In fact, male and female foxes are so similar that, after studying red foxes for years, I still have to look to see if a fox has a penis sheath to tell if it is a dog fox or vixen.

Other wild canid species, such as the wolf and coyote, show greater sexual dimorphism. Why not the red fox? As mentioned above the explanation seems to lie in their monogamy, with the real reason resting with the vixen. Once the foxes have formed their pair bond, it is not in the vixen's interest to prod the male to prove himself in further male–male contests. If the dog fox is injured as a result of this bravado, she risks the chance that her whole litter may starve. In carnivore species that are more promiscuous and in which the male does not help to raise the young, such as bears and weasels, it is in the female's genetic interest to have the males compete and fight over her in as many contests as possible. This is how she will get the best genetic stock to father her offspring. In these species the males are often larger and stronger than the females, and they roam extensively during the mating season and compete actively to breed as many females as possible.

These aspects about the red fox's livelihood and family life are summarized in the following diagram:

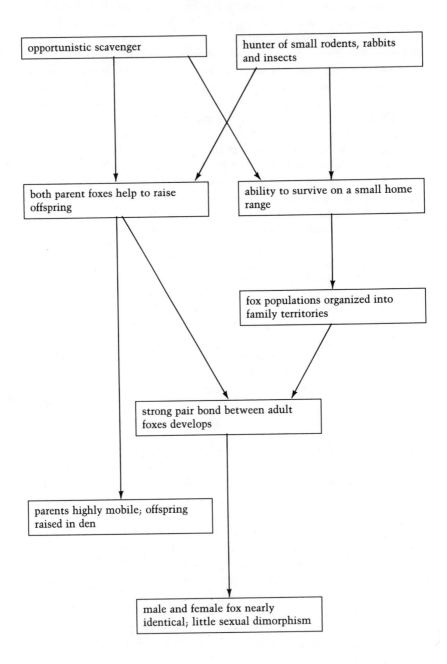

A vixen will nurse her kits for five weeks.

Adulthood

Now let us take a look at how young red foxes develop into
adults. The defended family territory continues to play an important role in the pups' later development, allowing them to make
the transition to adulthood in relatively safe and secure surroundings. Typically the parent foxes continue to bring food to the pups
at the den until they are about three-and-a-half months old. Then
the kits begin to move around on the family territory and find
food on their own. At first they may only be able to capture
insects and eat wild fruit, and the parents must still provide part
of their food, but gradually the kits become more competent as
predators. This slow transition to independence continues until
the young foxes are about seven months old, but at this point
peace in the family begins to erode. During September and October, competitive behavior (or at least gonadal activity that leads to
changes in male sex hormones) in both the dog fox and male
offspring stresses their relationship until the young males leave

the family territory and set out to search for a mate and vacant territory of their own.

The fate of the young female offspring is more complicated. Young vixens may also disperse from the family territory, but they usually do so in late autumn or early winter, a month or so after the young males have left. On the other hand, if conditions on the family territory are favorable, one—or in rare cases several—of the young vixens may stay on the territory for up to several years. The mother and daughters now form a strict dominance hierarchy with the older vixen in the top position. The next spring the male fox and his mate breed, and all the young vixens help in raising their pups. In most years these subordinate vixens do not have litters of their own but only serve as "helpers" in raising their parents' subsequent litters. We shall explore this cooperative relationship between mother and daughters in greater detail in the next chapter.

These facets about the fox's family life are summarized below:

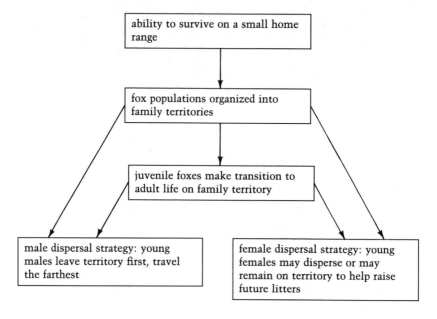

Fox and Man

One more piece of the fox's life needs to be fitted into place—its relationship to man. From what has already been said, this relationship can be easily understood. Foxes prefer varied, semiopen country. This fact helps to explain why red foxes often prosper in rural areas despite bounties, poison, trapping, hunting, and other

A juvenile fox stalks its prey.

forms of destruction inflicted on them. Many of man's activities create openings and edges in the landscape. Forests are cleared for lumber and to create agricultural fields, orchards, or to make room for buildings. Closer to the city, man develops parks and golf courses—grassy areas bordered by woodlots. Even roads that cut through forests create narrow openings with brush borders. These landscape alterations produce shrubby edges that inadvertently create prime hunting terrain for the red fox. Furthermore, there is often enough heavily shrubbed or forested land nearby so that foxes can escape from their enemies as well as find suitable denning habitat. As a result, red foxes are prospering and expanding their range into new areas as man continues to eliminate some of the fox's competitors such as the coyote, bobcat, and wolf. Despite man's often vigorous attempts to control them, there are probably more red foxes alive in the world today than ever before.

We have come about as far as we can in assembling the fox's life. The following diagram brings together many of its interlocking components. Each individual piece of the fox's life flows from the part that went before and leads to the part that follows. The end result is an intricate whole, a life strategy that has supported the species during the eons of its existence. I have pointed out in this effort only the primary ways the various facets of the fox's

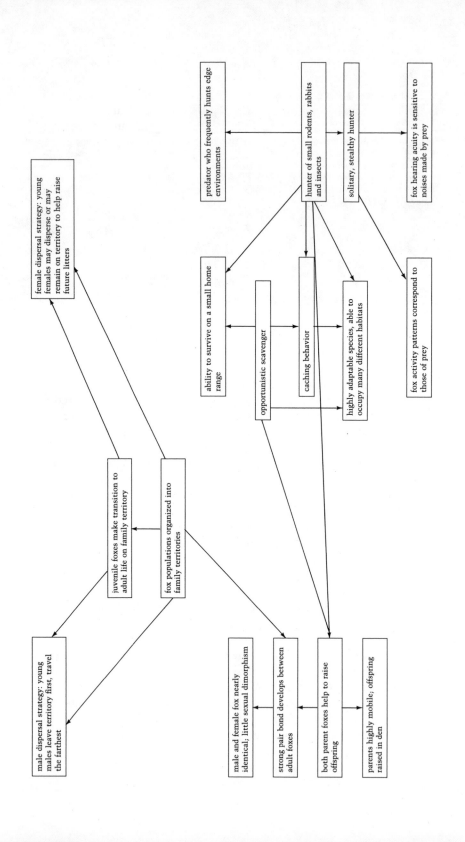

life join together. Other relationships exist, and new ones are developing so that more arrows and new boxes could be added to the diagram. For example, the fox's scavenging also benefits from many of man's practices, whether the fox is robbing fruit from an orchard, scavenging compost piles, or harvesting earthworms from lawns at night. Other relationships are possible, but one must stop somewhere.

We have only begun to assemble the fox's life. To understand the fox further, we must leave our flow charts and actually follow foxes and watch what they do. To say that foxes cache does not do justice to the ploys and counterploys that the fox has evolved to protect its hoarded food. To say that the fox is a predator of mice, rabbits, and insects does not describe the intriguing strategies the fox uses to hunt each of its prey. We have an overview of how the elements of a fox's life fit together, a skeletal understanding. Now let us flesh it out by following and observing freeranging foxes.

3 *Growing Up Fox Style*

It happens during the coldest part of the year—a time when Northerners bundle up against twilight days and -40 degree weather with parkas and windpants, gauntlets and mukluks. It is a time when darkness pervades the forest, when everything seems brittle.

Yet during this inhospitable period of late January or early February red foxes begin to court—anticipating the promise of spring at a time when most animals are simply surviving the frozen wastes of a boreal winter. It is not difficult to know when the foxes are feeling amorous; a musky odor begins to emanate from the urine stains they leave in the snow. In my area both male and female foxes during the span of just a week or so develop this breeding odor, and the scent is particularly strong in the urine marks of the males. Then for the next month or six weeks a skunky fragrance permeates the woods whenever a red fox and his vixen have established their territory, that is, the area they use and defend against other foxes. After studying foxes for years, I can say with confidence that it is my nose more than anything that first tells me when foxes are courting.

During this amatory interlude, the dog fox and vixen, both of whom have been leading solitary lives on the territory, now begin to seek one another out. For several weeks they travel together—for at least part of their time—and become increasingly at ease with being close and with touching one another, until they breed. During the courtship period, the vixen also visits and cleans out several lairs on the territory, one of which she will choose as her whelping den.

Why does a vixen choose to give birth to her pups in one place rather than another? What does she see in a particular piece of earth? After surveying 35 whelping dens, I found that they have certain typical characteristics. The den is usually located on a hillside in sandy loam. It is often in the forest but close to a meadow or open slope. And it usually has multiple entrances, the largest one about ten inches in diameter. Finally the den retreat is normally within 100 yards of a water source, although this source can be anything from a large lake to a humble pool in a muskeg. With these requirements vixens have to be selective about where they bear and raise their cubs.

Certain of these features, such as the nearness of drinking water, have obvious advantages; but I am not sure I understand the others. Why are they so common? After reflecting about them, I realized that vixens probably select sandy loam because they can dig into it more easily. But it could also be that this soil, combined with the hillside location and nearby open area, may make for a site where snow melts early, frost leaves the ground quickly, and drainage is always good. Vixens use the surrounding forest for shelter and escape; they can sun themselves in the adjacent meadow, and their rambunctious pups can play freely there.

A vixen may use the same whelping den year after year, and when she dies the site may be passed on to one of her daughters. One fox lair near my cabin, for instance, has been active for nine of the last twelve years. Because adult foxes in the wild usually

During January the normally solitary tracks of a red fox become paired as courtship begins.

live only three to seven years, this den probably has been occupied by several different fox families. On the other hand, abandoned lairs that deteriorate through disuse may be renovated after a long period of vacancy. The first den I made detailed observations about in 1972 was active until 1974. After that it lay vacant until a young fox couple reoccupied it and raised a family there during the spring of 1980.

As extra protection foxes never seem to rely on just the whelping den, but always have other smaller burrows hidden away in secluded spots on their territory. Thus if they are disturbed or threatened at the whelping site, the vixen or dog fox will simply move the pups to a new location. Frequently the vixen initiates the move, and it sometimes leaves her mate in a quandary.

I remember one such experience when, for an extended period, I was observing foxes at a den 24 hours a day to document how six-week-old pups were being raised. One afternoon I moved in close, attempting to get some photographs; but unwittingly I had crossed an invisible but important vulpine boundary, and the vixen acted as if she had decided enough was enough. She waited until the middle of the night and then moved the pups, not carrying them like other canids but marching them along behind her. About first light the dog fox returned to the den with a rabbit in his mouth. He went to the main entrance and called the pups out with the rapid "wuk . . . wuk . . . wuk" chortle that foxes use, but no one came forth. Setting the rabbit down, he began to smell around the area. He appeared to be able to figure out what had happened by the odors that remained on the ground. Sniffing the site thoroughly, he finally detected the trail where the vixen and four kits had traveled off through the forest. He then returned, cached the rabbit, and began to track down his family, following their scent through the bush. On the other hand, I, not being blessed with the nose of a fox, spent the next seven days searching in vain for the new den site.

But I am getting ahead of myself. Let me backtrack and follow the pups through their eventful early life. After a gestation period of approximately 52 days, the kits are born. In my area this occurs during late March or early April. The litter on the average contains five whining, hungry mouths; but a litter of nine cubs is not uncommon. Regardless of the size of the litter at birth, frequently some cubs die. The number that survive depends among other things upon the amount of food the adult foxes can find on their territory.

When the kits are born, they are helpless and weigh less than a quarter of a pound. Except for their smaller size, they are remarkably similar to newborn coyote and wolf pups—with one impor-

46

A vixen close to term rests by the den awaiting the birth of her pups.

tant exception. Even at birth most young foxes have a curious white tip on the tail. Red foxes retain this white tag throughout their life, and they are the only North American canid species to show this distinctive marking.

During their first month of life, young foxes do not possess the brilliant red coat that is the hallmark of the species; instead their coats are charcoal gray. Foxes are not alone in this; at birth the whelps of coyotes, wolves, and other species of wild canids are also cloaked with dark gray fur. As far as I can determine, this

47

special natal fur does not have any physiological function. It may serve as camouflage while the young canids remain within the darkness of the den, but I believe there is more to it than that. It is noteworthy that other families of carnivores who also raise their young in burrows produce whelps with special birth coats, but ones quite different from the dark gray of young canids. For example, many species of weasels (Mustelidae) have kits covered with white or cream-colored fur. Wild cat species also give birth to young that exhibit spotted or striped infantile fur. Consequently these groups of carnivores have offspring cloaked in distinctively different coats. I believe this natal pelage may be an important cue for the adult carnivores. Ethologist Konrad Lorenz and more recently evolutionary biologist Stephen Jay Gould suggest that baby features commonly observed in human infants— such as a head large in proportion to the body, protruding forehead, small nose, rounded body shape, and short, thick extremities—elicit caring behaviors and affectionate responses from parents and other adults. Lorenz further suggests that these features are the reasons we find baby animals, teddy bears, and toy dogs, such as the Pekinese, cute and cuddly; they all exhibit these endearing characteristics. But the infantile features mentioned above are not unique to human infants; the young of many mammal species exhibit these qualities. It is interesting to suggest that the mechanism that Lorenz and Gould discussed for humans may also be working in other species of mammals. These infantile characteristics help to endear the young animals to their parents and help to elicit from them caring and protective behaviors. I further speculate that for a number of carnivore species a special natal coat has evolved that may add to this effect. Specifically these distinctive birth coats may also help elicit parental behaviors and caring responses from adult carnivores and may have evolved to help establish a strong emotional bond between parent and offspring.

Fox kits grow slowly during the first few weeks, and their eyes do not open until they are ten or twelve days old. In many southern areas, the vixen may leave the den after two days, but in northern Saskatchewan the vixen constantly attends her pups. From a few days prior to birth until the kits are approximately ten days old, she is constantly with them in the den and completely dependent on the male for her food, which he provides with regularity.

Why is the vixen unwilling to leave the kits? David Macdonald, who has studied red foxes extensively in Great Britain, suggests that the kits are so small and vulnerable during early life that the vixen must surround them with her warm body if they are to survive. This maternal protection becomes increasingly important the farther north one goes. When kits are born, the ground in northern Saskatchewan is still frozen and covered with snow. The den is damp, and the air temperature inside is a chilly zero degrees Centigrade or colder. The vixen acts as a thermal blanket, protecting the cubs in their frosty underground chamber. One could almost say she incubates them.

The vixen begins to leave the den and resume her normal travels when her cubs are approximately two weeks old. By then they are well furred and large enough to maintain their own body temperature. The vixen now spends more and more time away from the lair hunting prey for herself, but she continues to receive some food from her mate. She does, however, return to the den at regular intervals to nurse the kits. While in the den, she attends to her other maternal duties as well—she plays or naps with the cubs and then grooms them, cleaning out their ears, licking their groin regions, and eating their waste products. She maintains this routine until she begins to wean the pups during their fifth week.

The early life of a fox kit is not the carefree, leisurely existence that one might imagine. It is business from an early age. During their second week, the kits begin to grow faster, and their milk teeth begin to come in—canines first, followed by incisors, and finally premolars. During their third week, they begin to experiment with their teeth. They chew and suck meat that their parents have brought them, and they learn to enjoy the juices found in freshly killed rabbit. Now they enter an important phase of their development.

During this stage the fox kits are far from fully coordinated and have been above ground only for short periods. Yet at about 25 days of age, the kits begin to fight viciously; they clash with each other in short, serious, and sometimes fatal contests (although fatalities are rare). Fox kits do not act like cute, cuddly puppies such as those of the domestic dog; rather they have always seemed to me to have a slightly demonic character. They are tough month-old thugs, little street fighters who initiate fights and establish a strict dominance hierarchy during the following ten days. The alpha (or dominant) member of the litter establishes itself, and the hierarchical process continues all the way down to the omega animal.

Evidence suggests that these dominance relationships among the cubs are stable and have a great bearing on their survival. If the adults are bringing the kits only a limited supply of food, competition for it is keen. By establishing a hierarchy, the kits determine who can steal food from whom. The largest member of the litter, whether male or female, usually becomes the alpha

*During early life fox kits fight viciously until they have established a
dominance hierarchy.*

pup. Using either intimidation or direct attack, this animal steals
food from its litter mates. Each pup steals food from litter mates
below it in the hierarchy. When food on the family territory is
scarce, it is a blunt fact of a fox's life that the dominant pups get
a larger portion of the food and have the best chance of surviving
while smaller and submissive ones may perish. The runt of the
litter dies first, then the second lowest on the hierarchy, and so
on. It is a brutal process, but given the constraints of the hierarchy it ensures that each kit can maximize its own chances for
survival. Nor have the parent foxes ever evolved behaviors to
interfere with this sibling rivalry, since this competition results
in a litter that is both composed of the healthiest whelps and
pared down so that its size correctly matches the food resources
of the territory.

By early May the fox kits in my area begin to come above
ground for longer periods of time. At this point the hierarchy is
solidly established, and the aggressiveness of the cubs actually
diminishes. Gradually they become more social, playful, and puppylike. Approximately at this time people begin to see the young
foxes. How much has transpired in the lives of fox cubs before we
are even aware of their presence!

When a parent fox returns to the den during early May and

50

chortles for the pups, they stumble out—unsteady on their feet, blue-eyed, and still dark-coated. But in my area at about five weeks of age an interesting change takes place: kits begin to lose their dark natal coat and grow a new pelage. Once again the pelage is not the brilliant red of an adult fox but a sandy-colored juvenile coat that matches the sandy soil of the den site to an impressive degree. I believe that this coat may help to camouflage the pups around the den site and protect them from occasional predators such as hawks, owls, and coyotes.

Reflecting on the sandy-colored coat, I began to realize that this juvenile pelage might have come about from an interesting series of interactions. I can imagine it evolving something like this: at first vixens tended to dig their main dens in sandy or sandy-loam soils simply because it afforded good digging. But over time the fur of the kits slowly evolved to match the sandy color of the soil. Evolutionary theory states that if the sandy-colored pups have even a slightly better chance of surviving over time most fox kits will evolve this adaptive coloration. As this happened it created greater selection pressure for the vixen to locate her den in sandy-colored soil. Consequently the selection of the den by the vixen and the juvenile pelage of the kits are coadapted to protect the young during their vulnerable first month outside the den.

The young foxes' buff coat is actually a specially colored under-fur (the underwool of adult foxes is a dark gray). The kits exhibit this juvenile coat for approximately five weeks, but beginning in early June outer red guard hairs begin to grow through it. By the end of June, kits display the bright red coats normally associated with adult foxes.

When the kits are five weeks old, they begin the slow but steady transition that will launch them into adult life. The pups now begin eating solid food, and the vixen starts to wean them. Weaning fox pups is an uncomplicated procedure. When the kits try to nurse, the vixen rolls over on her stomach. If they persist she threatens them and runs off a short distance. She becomes increasingly unreceptive to their suckling, and by the time they are eight weeks old the pups are completely weaned. But being weaned doesn't mean being cut off. The parents continue to hunt and scavenge at different times of the day and night to supply the kits with sustenance. These adults bring in a veritable parade of prey and present it to their pups. For example, in one week I have seen the fox's menu include mice, voles, hares, squirrels, ground-hogs, songbirds, grouse, and a spawning long-nosed sucker.

Distributing this food is a household ritual that always occurs the same way. Carrying the item in its mouth, the adult fox arrives at the den and chortles, and one or more of the pups rush out to greet it. The first pup to reach the parent crouches low and beats its tail about wildly. The kit whines and creeps toward the adult. Then reaching up it smells, licks, and bites at the corners of the adult's mouth.

There seems to be a consistent rule in fox families that the adult gives the food to the first pup that begs for it. This appears to be the mechanism by which supplies are more or less evenly distributed. Pups that have not eaten for awhile often rest outside the den and keep a more attentive watch; they race for the parent, and the first one there is given the food. However, once the kit has its ration its problems really begin. It now has to defend its prize against its litter mates. The kit tries to do this either by running off or by threatening any sibling who comes near. As I explained earlier, a dominant pup will frequently challenge and steal the provision from its subordinant litter mate.

Food-begging behavior also appears many months later in a completely different context—as part of the submissive behaviors shown by adult foxes. When two full-grown animals meet, the subordinate fox often acts as if it were a young pup begging to be fed. It crouches low, whines, and beats its tail madly in every direction. This keeps the white tip of the tail in constant motion (much like waving a white handkerchief), letting the tail function as an attention-getting device. The subordinate fox slowly creeps up to the dominant fox and carefully reaches up to smell and lick the corner of its mouth. This adult use of derived infantile behavior is not unique among red foxes. Embracing among adults, which is used as a greeting and reassuring gesture not only among humans but also chimpanzees and baboons, may be derived from infantile grasping. People in love sometimes coo and baby-talk to each other; we sometimes stammer or show shy juvenile behaviors when introduced to famous or influential people. Human displays, it turns out, are not altogether different from fox displays.

The role of the dog fox in raising the kit has been much debated among naturalists and field biologists. Some state that the male fox is never seen around the den, while others conclude that he supplies the majority of food for the pups. There is considerable variation among red foxes of different areas, the causes of which are not fully understood. However, in Prince Albert National Park, an area where the foxes are almost completely protected from hunting and trapping, I have consistently observed male foxes bringing food to the pups until they are approximately ten weeks old.

As I have already mentioned, in addition to the parents other

The vixen returns every few hours to nurse her pups.

foxes are occasionally observed bringing prey to the pups. These "helper" foxes are an interesting phenomenon. Most often the assistant appears to be a daughter from the previous year's litter who has stayed on the family territory but who has not given birth to pups of her own. When she is present, she makes a definite contribution to rearing the pups.

This helper illustrates an important point about evolution. Evolutionary theory states that individuals of a species that are most physically and behaviorally fit for their environment will leave the greatest number of offspring. But 20 years ago, British biologist W.D. Hamilton stressed that an individual's genetic relationships are also important. For example, red fox helpers are as closely related to their full sibling as they would be to their own offspring. If an assistant increases the survival rate of her younger brothers and sisters, she helps some of her own genes survive into future generations.

Foxes that aid in raising their younger brothers and sisters may also derive other benefits. For example, they may gain valuable experience, which in the future will help them to rear their own offspring more successfully. In a few cases, they may also inherit part of their parents' territory.

Hamilton's theory makes another prediction about these young vixens. It states that if food is scarce the daughter will help raise and feed her mother's litter. If food is abundant, however, it predicts that the male fox will breed both the older vixen and his daughter, and both vixens will raise litters on the family territory. The male will bring food to both litters, and the vixens instead of competing will tolerate each other. Why does the theory predict these events? Because when food is abundant—considering the genetic relationships involved—each fox will pass on more of his or her genes to the next generation if multiple litters are reared on the territory. Thus because it is in all the foxes' "genetic interest" we can expect these behaviors to evolve.

My field observations tentatively seem to support these predictions. In my study area, during years when prey was abundant (one year, for example, there was a peak in the snowshoe hare population) I have occasionally observed two vixens raising separate litters several hundred yards from each other and once within the same den. In both cases the vixens got along amicably, and one dog fox brought food to all the pups. These observations suggest that the vixens were acquainted with each other and may have been related. These preliminary data are interesting, but before we draw any final conclusions about this aspect of the fox's family life more detailed research is needed.

During middle and late June, the pups' transition to adult life

A seven-week-old pup begging food from its mother.

continues. By this time they have grown to be nearly two-thirds the size of the adults; they are fully coordinated and exhibit a bright red coat. During this time the parents begin to visit the den less frequently, and the supply of food they provide begins to diminish.

At this point, as with many species, a classic conflict between parent and offspring arises. The fox kits appear perfectly content to remain at the den, playing continuously and being well fed by their parents. But the adult foxes, who have been run ragged, seem to have some other arrangement in mind, and, fortunately for them, they are well in control of the situation.

During June the pups spend endless hours waiting for the adults to arrive with food, but the adults spend more and more time away from the den. When an adult appears, one pup gets the food, which he quickly consumes. Then the adult frequently takes one or several of the pups on exploratory trips away from the site. At

When two adult foxes meet, the subordinant animal often displays a puppylike begging posture.

first the parents lead the kits away and return with them. Later the kits return on their own. I wish I could have gone along, but the parent foxes never tolerated me on these expeditions so this aspect of fox life has always eluded me.

The kits, while hanging around the den, begin to eat wild fruit and to hunt insects and whatever other prey they can find. Gradually they become more confident, venture forth slightly farther each day, and spend less time at the lair. It seems that a combination of boredom and hunger eventually drives the kits away from the den site, whereupon they start a predatory, scavenging life of their own.

At first the pups travel in groups of two or three. But because their prey are small, quick, and must be hunted in a catlike manner, the young foxes are slowly forced to become solitary predators.

Occasionally the vixen and pups will rendezvous on the territory during late summer and early fall. When this happens the foxes greet, play, and generally enjoy each other. By this point, however, the dog fox is intolerant of his cubs. During September the male kits begin to mature sexually, and this increases the competition between father and son.

No one is certain what causes the kits to finally leave the territory altogether. Adult foxes have not been seen aggressively driving their young away, but they may exclude them from the best hunting and scavenging sites. Whatever the cause, during late September or October the kits begin to disperse. Males leave first and travel the farthest. Each tends to travel in one direction until he has found an available territory. Once male kits have left their birthplace, they usually do not return. Female kits may remain on the family territory into December, and if food is plentiful one or several female offspring may stay on.

Such is the way it happens. Each spring kits are born; as help-less as embryos they would freeze without their mother's constant attention. In the course of six short months, these tiny, helpless creatures grow into self-sufficient, graceful, flame-colored predators that leave their families and seek out their own part of the forest. When a young fox leaves, it becomes self-reliant and takes on the ways and manners of adults. It claims a territory and inhabits it with another fox—its prospective mate. During the frozen darkness of winter, the pair sets up its territory and courts, and the cycle inherent in a fox's life completes itself.

4   Fox Hunting

Thirty yards in front of me, the fox stopped suddenly. It turned broadside on the deer run, lifted its head, and perked it ears. Then slowly it stalked off the trail, testing each foothold, careful not to break a twig or rustle a dry leaf. Fixing its eyes on small movements in the grass in front of it, the fox froze, then coiled into a deep crouch. It waited, adjusted its feet, and lunged, catapulting itself over a low shrub. While in the air, it beat its plumelike tail several times to the left so as to turn its body slightly in that direction. The fox landed as gracefully as it had launched itself, its forepaws touching down where the grass had been moving moments before. The fox tried to pin a mouse to the ground, but the prey escaped into a burrow, one of many escape hatches that punctuate a mouse's home range. After a brief search, the fox returned to the deer run and continued to walk down it searching for other prey.

What I had just witnessed was by now a familiar sight to me. Every day for a month from dawn until dusk I had been observing red foxes as they eked out a living in the northern forest. I was ten pounds lighter now and finally fit enough to keep up with these lithe carnivores.

When a fox hunts, it shows signs of its predatory activities, for example, by the way it walks down a path searching for prey or by its increased alertness and excitedness. Whenever I saw signs that a fox was hunting, I dropped 50 to 100 yards behind and followed it at that distance. When a fox actually spotted a prey, I remained perfectly still, describing that hunt into a small field tape recorder. At this distance the foxes showed no detectable reaction to my presence as they carried out their predatory missions.

Collecting observations on these foxes would be an easy task if two conditions were met: the fox stayed on the trail and the trail was straight without forks or diversions. But these conditions are seldom encountered in the boreal forest. The paths and animal runs that foxes use for hunting twist and wind through the woods. Furthermore, as soon as the fox disappears behind a shrub or around a corner or steps off the trail into dense brush, visual contact must be reestablished within a fraction of a minute or the subject becomes lost and the results of hours of hard work are truncated. I am still amazed at how quickly foxes can disappear into the woods—like a puff of campfire smoke disappearing into the canopy of trees. In a few seconds the fox can vanish from view.

This problem arises partly because the foxes I study have only a minor incentive for allowing me to follow them. In my field studies I give the foxes a small piece of meat when we find each other at the start of an observation session. This tasty morsel, which the foxes have come to expect, is my telemetry system; it is how I locate foxes on my study area. After this initial reward, I want them to go about their activities with my presence influencing their behavior as little as possible. Over a week or ten days, I teach the foxes that after receiving their initial snack they will get no more food; they set about their activities, and I slowly teach them to accept me. It is therefore entirely up to me to maintain visual contact with the animals. Keeping up to a fox on the hunt is not an easy job, especially while finding your footing over rough terrain, recording fieldnotes into a small tape recorder, and managing cameras and binoculars—all frequently executed at a full run. I found that it usually took a month to become fit and rugged enough to keep up with these animals. It also took the foxes about that long to get used to my following them. As a result only after an important month of conditioning for both the foxes and myself were we ready to document how red foxes hunt their prey in the pristine boreal environment of Prince Albert National Park.

Carnivorous Patterns

Before considering the details of my field study, let me put my findings in a broader context. My fox research is only one of a number of field studies completed on freeranging carnivores during the course of the past several decades. Wolves, lions, hyenas, mountain lions, coyotes, and lynx are just some of the predators that have been studied. When one compares the results of this research, some fascinating patterns emerge about how animals hunt. An understanding of these patterns also clarifies many aspects of the hunting behavior of the red fox.

The studies show that members of the cat family (felids) and members of the dog family (canids) use distinctly different hunting strategies. Felids are generally built for stealth and bursts of speed; canids for endurance and distance running. Felids stalk close and make surprise attacks. Canids approach prey openly, outpace it during a long run, and finally pull it down. Consequently the hunting adaptations of felids are different from those of canids. Cats have evolved soft foot pads for stalking, a fine sense of balance for a motionless posture during the stalk, and hooked retractile claws and great coordination in the forelimbs for capturing prey. To make their killing bite effective, cats have evolved short muzzles, strong jaw muscles, and daggerlike canines.

Dogs possess different equipment for hunting. They have evolved adaptations for speed and stamina that help them in cursorial hunting. In addition, certain canids show highly developed social behaviors that allow packs of these hunters to overpower and kill large animals. Predation by canids generally is not based on stealth but on the detection, testing, and running down of weak or vulnerable animals. Dog packs also dispatch their prey in a manner different from cats. They kill by repeated biting, mainly at the prey's hind end. When the prey is weakened from these wounds, the pack eviscerates the animal and commences feeding.

There are interesting exceptions to these general evolutionary patterns. For example, the cheetah is a cat that hunts in a doglike manner. On the African savanna it approaches prey openly, pursues it in a fast run, and finally knocks it over. As a result the cheetah shows both structural and behavioral convergent evolution with canids: it has evolved a slightly elongated muzzle; long legs and a deep chest; and blunt claws with thick, hard foot pads—all adaptations for speed and capturing prey on the run.

The lion is another exception because it practices communal hunting. George Schaller, who studied lions on the Serengeti for four years, calls the lion "the most sociable cat." In this aspect it resembles the wolf and African hunting dog. The adaptive value of communal hunting seems to be the same for all three predators, namely, more efficient searching for prey plus the abilities to capture large prey and defend the carcass against other predators that may try to steal it.

The lion and the cheetah show that ecological selection pressures can cause a species to deviate from the hunting strategies typically shown by its taxonomic family and evolve hunting strategies similar to another taxonomic group. When I began studying the red fox, I wondered where in these patterns the fox would fit. Basically I set out to find out what kind of predator the red fox is.

The Fox as Hunter

I began to understand the fox as predator by looking at the type of quarry it catches. A review of food habits shows that red foxes catch mainly small animals—mice, voles, songbirds, rabbits, and insects such as beetles and grasshoppers. A red fox may tackle a prey as large as a grouse, ptarmigan, or jackrabbit, but larger animals, such as deer or sheep, are usually scavenged upon rather than killed by red foxes.

In essence, the red fox is a slayer of small prey. In the course of my research, I observed red foxes hunting 34 different kinds of animals—all of them small. A roster of these prey follows:

Class	Popular Name	Scientific Name
Mammals	Meadow vole	*Microtus pennsylvanicus*
	Boreal redback vole	*Clethrionomys gapperi*
	Deer mouse	*Peromyscus maniculatus*
	Red squirrel	*Tamiasciurus hudsonicus*
	Snowshoe hare	*Lepus americanus*
	Masked shrew	*Sorex cinereus*
	Arctic shrew	*Sorex arcticus*
	Northern water shrew	*Sorex palustris*
	Northern pocket gopher	*Thomomys talpoides*
	Woodchuck	*Marmota monax*
	Least chipmunk	*Eutamias minimus*
	Northern bog lemming	*Synaptomys borealis*
	Meadow jumping mouse	*Zapus hudsonius*
	Muskrat	*Ondatra zibethica*
Birds	Dark-eyed junco	*Junco hyemalis*
	Song sparrow	*Melospiza melodia*
	White-crowned sparrow	*Zonotrichia leucophrys*
	White-throated sparrow	*Zonotrichia albicollis*
	Tree sparrow	*Spizella arborea*
	Snow bunting	*Plectrophenax nivalis*
	Black-capped chickadee	*Parus atricapillus*
	Boreal chickadee	*Parus hudsonicus*
	Gray jay	*Perisoreus canadensis*
	Black-billed magpie	*Pica pica*
	Ruffed grouse	*Bonasa umbellus*
	Spruce grouse	*Dendragapus canadensis*
	Mallard	*Anas platyrhynchos*
Insects	Crickets and grasshoppers	Orthoptera
	Giant water beetle	*Benacus griseus*
	Giant water beetle	*Lethocerus americanus*
	Beetles	Coleoptera
	Moths	Lepidoptera
	Flies	Diptera

A snowshoe hare absorbs solar warmth on a cold winter's day. Hares are among the largest animals hunted by the red fox.

Within each class of animal (mammal, bird, and insect), I have arranged the species so that the ones I observed the foxes hunting most are given first and the ones that were hunted rarely are given last. The frequency with which foxes hunt a particular species is probably influenced by many things—the animal's abundance, the fox's preference for it, and the fox's ability to detect and capture that prey. Furthermore, the "huntability" of an animal may change with different seasons and different environmental conditions. The above list aims at giving a general idea of which prey were hunted the most and which were hunted the least by the foxes on my study area.

From previous studies, particularly those carried out by ethologists Gunter Tembrock and Michael Fox, I knew that the red fox usually captures prey either by biting it or by pinning it to the ground with a vigorous forepaw stab. In my field studies, one of these capture attempts had to be made before the behavior was termed a "completed hunt." If a sequence terminated before a fox made a biting attempt or a forepaw stab at the prey, the sequence was classified "searching after prey" rather than a "hunt." The data for this study consist of 434 completed hunts observed in 22 different, freeranging red foxes; 139 of which (32 percent) were successful. It is interesting to compare how successful the foxes were at catching various prey. In the hunts I observed, the foxes captured mammals 23 percent of the time (60 out of 257 hunts were successful); they caught birds a little more than 2 percent of the time (2 out of 83 hunts), and they caught insects 82 percent of the time (77 out of 94 hunts). These differences in success rate are explained by the effectiveness with which the prey escaped. For example, grasshoppers and crickets, the insects hunted most by the foxes, showed feckless escape behaviors. If the fox missed one of these prey, the insect often hopped only one or two strides away; the fox watched where it landed and was often able to relocate and capture it. On the other hand, mammals frequently escaped the jaws of the fox by either outrunning the predator, fleeing down a burrow, or scurrying under a boulder or log. Red squirrels also escaped by climbing trees. Small birds foiled the fox more successfully than any other prey; I observed a fox capturing a small bird only once in 71 hunts. Passerine prey have extremely quick reactions, and they also alternate feeding on the ground with periodically flying up to perches where they are often able to spot a predator stalking them. From the fox's capture record, it appears that passerines may have the hardest set of defenses for a fox to circumvent successfully. On the other hand, Hans Kruuk in Great Britain and Alan Sargeant in the United States have documented that if conditions happen to be right ground-nesting birds, such as certain species of ducks and gulls, can be quite vulnerable to foxes during the spring when they are incubating eggs. At this time the avian content in a fox's diet may increase markedly, but no biologist that I know of has found birds to form a significant portion of the fox's year-round diet.

What kind of taxonomist is the red fox? How does it "perceive" its prey? Does it hunt mammals in one manner, birds differently, and insects with a third hunting strategy? If not, how does the fox respond to its quarry? These questions pose a challenging problem in fox perception, and the only clues we can glean come from the fox's hunting behavior. After analyzing more than 400 hunts, it became clear to me that foxes hunt prey according to a taxonomy slightly different from our own. The foxes I observed exhibited several distinct hunting strategies, with each strategy directed almost exclusively at different prey.

The first prey group consists of the mammals in the above list minus the red squirrel and snowshoe hare. At first glance this group appears to consist of rodents, but it also contains some insectivores (shrews). After watching the foxes hunt these mammals, I became convinced that what these prey have in common, at least from the fox's point of view, is that they escape from predators by fleeing down a burrow or dashing under an object such as a rock or fallen tree. Consequently I labeled this group of prey "small burrowing mammals."

The second prey group includes rabbits and hares. Lagomorphs, as biologists call members of the rabbit family, often flee from the fox by running, and the strategy foxes employ for hunting these fleet-footed prey reflects this fact.

The third prey group consists of birds and tree squirrels. These prey have in common: (1) high visual acuity, (2) they often flee upward to escape from the fox, and (3) they frequently keep a watch for predators from high perches. These attributes will be used to explain certain features of the hunting strategy that foxes use to capture arboreal prey. The fourth prey group consists of insects—prey that the fox hunts in a more casual manner.

This division of prey was not principally my idea. I did not set out into the field with these groupings in mind; rather the foxes "dictated" them to me through their hunting behavior. Foxes on my study area responded to these four prey groups differently, usually hunting each with a distinct tactic. Let us begin to survey the fox's hunting arsenal by looking at how foxes capture prey from this last group—the insects.

Insects

The foxes I studied usually captured ground insects (crickets, grasshoppers, and beetles) but occasionally flies and moths. When foxes hunt insects, they rarely exhibit the high degree of tenseness and excitedness that they do when hunting other types of prey. The fox's casualness may be explained by both the weakly developed escape behaviors that many insects show and the fact that individual insects only provide the fox with tidbits of food. A typical hunt taken from my fieldnotes reads:

> Hunt No. 128. 2 October 1971. 11:47 a.m.
> This hunt was a successful grasshopper
> hunt. My Friend (a young male fox) was
> walking over to his sunning spot when he
> apparently heard something in the grass. He
> became alert; he crouched his shoulders
> slightly, lowered his head and neck, and
> peered at the spot. He walked a few paces
> over to where the grasshopper was and began

to smell for the prey, apparently having lost sight of it. The grasshopper jumped, and the fox bit but missed. The grasshopper jumped again, sailing about four feet away; My Friend watched closely, and when the prey landed he walked over, bit, and got the prey. Then the fox lifted his head and consumed the prey. After that he walked over to his sunning spot and lay down.

This incident shows many characteristics typical of an insect hunt. Usually a fox is traveling along, scavenging, cleaning out a den, or doing some other activity when it happens to detect an insect in nearby grasses or in the leaves on the forest floor. Then the fox stops what it is doing, walks over, bites down, and chews up the prey. Many times it is as simple as that. On other occasions, however, the insect hops or flies away, and the fox must search for it by poking its muzzle into shrubs or stamping one of its forefeet on the forest floor where it suspects the insect might be. After prodding the insect to move, the fox tries to locate it from small movements or rustling sounds. David Macdonald in Great Britain reports red foxes using foot stamping when they are hunting earthworms, but I have never observed them to use it in

Searching for small mammals, this red fox raises its head and neck into the characteristic "mousing" position.

any other kind of hunt. Nor have I ever observed a red fox to use a foot-stamping action to crush a prey. The casual demeanor that foxes exhibit while capturing insects is unique to this type of hunting; the fox's other prey offer more challenging contests.

Small Burrowing Mammals

The fox's strategies for capturing chipmunks, pocket gophers, and other burrowing mammals are considerably more complex than those used for capturing insects. In the boreal forest, foxes usually hunt small burrowing mammals while traveling along game trails, rabbit paths, hiking trails, or the edge of roadsides. The foxes hunt from these pathways apparently for two reasons. First, if the trail is wide enough to let in sunlight dense vegetation grows adjacent to it and small mammals are frequently abundant. Second, foxes appear to use trails for quietness. The fox must approach these prey without being detected. If the fox makes a sound, such as cracking a dry twig, the prey will either crouch motionless making detection difficult or flee down a burrow or under a log.

The hunt begins with the fox walking down a trail. When it suspects a prey is adjacent to the path, the fox turns perpendicular to the trail, lifts its head and neck high, and stares intently. This elevated head posture is so characteristic of small borrowing mammal hunts that I termed it "head and neck in mousing position." A fox showing this head posture is almost always hunting.

If the fox cannot locate the prey, it may lower its head and sniff, trying to locate the prey from its scent. Or the fox may lift its head and rotate it, cocking it first one way and then another. These small head movements place the fox's ears at slightly different distances and angles from the sound. As mentioned earlier, Henrik Österholm has shown that captive red foxes can locate a small rustling sound within one degree of its true location.

Alternatively the fox may search for prey by stalking. When stalking to get closer to the quarry, the fox lowers its head and crouches down slightly. It moves slowly, carefully putting a forefoot down lightly on a spot and occasionally bringing it back up and choosing again. It tries to make as little noise as possible. To further reduce noise, the fox puts its hindfeet exactly where its forefeet had been.

After stalking for several yards, the fox assumes head in mousing position, and when it has located the prey it crouches deeply and then lunges, arcing through the air. At the end of the lunge, the fox tries to pin the prey to the ground. Then it may dispatch the prey with several quick bites to the body or carry it off to a safe area where it plays with the prey like a cat with a mouse.

The fox's lunges are normally short (2 to 6 feet), but they can also be impressively long. I have observed red foxes catapult 15 feet through the air from a standing start to capture prey. By way of comparison, the coyote, which is a third or more larger than the fox, has been observed to perform hunting lunges that are only 8 feet or less in length. Foxes also show a preference for hunting the downhill side of trails. They may do this because by lunging downhill more area is placed within their reach. After observing several impressive downhill lunges, I made careful measurements and found that these foxes were airborne for a distance of up to 25 feet.

Arboreal Prey

When hunting birds and tree squirrels, red foxes rarely lunge; instead they stalk close and then charge the prey in a crouched, dashing run. As mentioned earlier the foxes I studied were not very successful in capturing birds; only slightly more than 2 percent of the hunts provided a meal. The foxes were equally unsuccessful in capturing red squirrels; only one out of forty-three hunts resulted in a capture. However, during spring I did observe adult foxes carrying freshly killed red squirrels to their pups at the den on four or five occasions.

The lack of success with which foxes hunt arboreal prey seems

To capture arboreal prey, the fox makes a quick, horizontal thrust rather than an arching lunge.

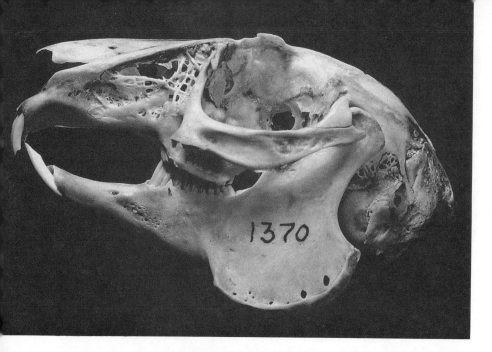

The unusually porous, fenestrated skull of a snowshoe hare may explain why red foxes are able to dispatch these prey quickly and easily.

to influence the way in which foxes hunt them. For example, on my study area foxes do not search for birds or tree squirrels, as they do for small burrowing mammals, but rather hunt them whenever they encounter them on the ground in vulnerable positions.

When a fox spots a bird or squirrel, it immediately crouches so low that its belly almost touches the ground. Its neck is held horizontal, ears alert, and it remains motionless while staring intently at its quarry. Then the fox "slink runs" at the prey, starting with a stalking pace, changing into a trot, and finally breaking into a gallop. The fox maintains its low, crouched posture throughout this charge. The last part of the slink run is a horizontal thrust jump. It ends with the fox trying to catch the prey with a bite. The forepaw stab that foxes use to capture small burrowing mammals is not used to capture birds or squirrels.

A fox will sometimes hunt tree squirrels by stalking them in a curious stop-and-start manner. When stalking a squirrel in this way, the fox carefully watches the squirrel and tries to synchronize its movements with the prey's. Each time the squirrel faces away from the fox and begins to forage, the fox takes several quick steps in a deeply crouched posture but freezes the instant the squirrel stops moving. The net effect is that the fox can mask its movements by synchronizing them to the movements of the squirrel. The fox appears tense and catlike when performing this synchronized stalking.

There is an interesting comparison between this type of stalking and the way a red fox stalks small burrowing mammals. The emphasis is different in each case. When stalking small burrowing mammals, the fox puts its feet down slowly and is careful not to make any noise, but it moves its head about freely, alternately searching for a quiet route and then looking at the place where it suspects to find the prey. On the other hand, during bird and squirrel hunts the fox may make little rustling sounds as its feet move across the ground, but it is always in a deeply crouched posture, and its head and eyes are fixed on the prey as though tied to it by an invisible cord. These two stalking techniques appear designed to exploit different weaknesses in the antipredator defenses of different prey. In small mammal hunts, the fox's stalking minimizes auditory cues that might alert the prey. During bird and squirrel hunts, stalking minimizes visual cues, while small noises are masked by the synchronization of the fox's movements with those of the prey.

Lagomorphs

Red foxes hunt rabbits and hares by stalking, followed by a "hell bent for leather" pursuit. This is the fourth hunting strategy that boreal red foxes show. In a typical hunt of a snowshoe hare, the fox crouches low to the ground and stalks slowly toward the prey while staring at it intently. Detecting the fox, the hare suddenly bolts and flees, often through thick cover, while the fox follows in a bounding gallop. When it gets close to the hare, the fox attempts to bite it in the leg or rump. Frequently at this point the hare will suddenly change direction. By zigzagging it tries to increase the distance between itself and the fox. These actions may be repeated, and the chase may cover a considerable amount of ground: the fox continues to come close, and the hare zigzags and runs off in a new direction. Finally the hare either escapes into a briar patch dense enough that the fox cannot follow or the predator gets a bite-hold on the prey. If the latter happens, the fox quickly pulls the hare off its feet and both collapse to the ground. The fox then quickly stands on its quarry, pins it to the ground with one or both forepaws, and bites it in the neck or head. Foxes are able to dispatch hares and rabbits quickly, perhaps because of the prey's slender neck and fenestrated skull.

Evolving the Catlike Canid

There can be little doubt that the red fox is a bona fide member of the family Canidae. It shows all the diagnostic morphological characteristics of this family including four well-developed toes; a digitigrade stance; semirigid, elongated legs; interlocking radius and ulna that prevents rotation of the front leg; no entepicondylar foramen of the humerus; a small, cartilagenous clavicle; a well-developed, grooved baculum; an elongated skull and zygomatic processes that project strongly outward; a smooth tongue; and 42 teeth with well-developed carnassials. Furthermore, it shows behavioral traits that are typical of Canidae. These include well-developed caching behavior, extensive scavenging behavior, hunting strategies that consist of lunging at small burrowing mammals and chasing lagomorphs over considerable distances, a well-developed pair bond, a social organization in which both parents help to raise the young, and burrow excavation behavior with the concomitant use of these burrows for raising young and providing shelter during storms. Cats, on the other hand, as shown by Devra Kleiman and John Eisenberg and others, exhibit none of these behavioral traits or show them only to a very limited degree. Furthermore, most cats show a truncated skull and foreshortened rostrum; they frequently cover their feces with litter and dirt, and they rotate their forearms and grasp prey with well-developed retractile claws. These are morphological and behavioral features not evident in the red fox. Consequently the taxonomic affinity of the red fox is clear: it is a bona fide member of the family Canidae.

Yet many aspects of the fox's hunting behavior are so feline that I was struck by the possibility of convergent evolution between foxes and cats. Therefore, I conducted a search for features shared by red foxes and various cats. This produced interesting results and led me to a new hypothesis concerning fox evolution.

Foxes and cats share numerous features. For example, many fox species have long, catlike vibrissae, or whiskers, on both the muzzle and the wrist (carpal joints). These vibrissae are proportionately longer in foxes than in other canid species. Field observations suggest that these whiskers function as tactile organs. The muzzle vibrissae may help to guide the fox's capture and killing bites, and the carpal vibrissae may assist the fox during stalking.

Foxes also exhibit proportionately long, thin but robust canine teeth that resemble the long, daggerlike canine teeth commonly observed in cats. Many predators once they have a small prey in their jaws shake their head vigorously to immobilize and kill the prey. But red foxes and cats do not often show this behavior. Instead they press their long canine teeth into the prey and keep exerting pressure until they damage the prey's central nervous system.

Red foxes also have felinelike paws. For example, red foxes can flex and partially retract their front claws so that they are in fact semiretractile. The toe and foot pads of the fox are small, and the rest of the foot is soft and covered with hair. I believe that these features not only reduce heat loss during winter but also function as adaptations for stalking by making the fox's feet more touch-sensitive.

Convergent evolution between the fox and cat, however, is most striking in the anatomy of their eyes. Both animals exhibit the combination of a vertical-slit pupil and a highly developed tapetum lucidum. The tapetum lucidum is a glistening layer of connective tissue that has evolved on the innermost sheath of the eyeball. It reflects light back out of the eye and causes the eyes of foxes and cats to occasionally glow a luminous dull green even though no strong light is shining into them. This glistening membrane acts like a mirror behind the retina so that light passes over the retina twice instead of once. Using analogous photographic terms, the tapetum lucidum and vertical-slit pupil represent the

A red fox's canine teeth are slender and almost daggerlike in appearance.

Unlike other canids foxes exhibit a vertical-slit pupil, an important adaptation that allows their highly developed night vision to accommodate to bright sunlight.

evolution of a light-multiplying device with a shutter that can stop way down to accommodate bright sunlight.

Foxes also show catlike features in their behaviors. For example, both red and gray foxes show a lateral threat display. When a fox threatens another fox with this display, it stands broadside, arches its back, erects its fur, and then charges broadside at the opponent in a stiff-legged gait. However, it is during hunting that the actions of a red fox most strongly resemble those of cats. For example, when hunting birds and squirrels, the fox stalks with its belly almost touching the ground or slink runs from cover to cover, crouching and waiting for its chance to ambush the prey.

What explanation is there for these similarities between foxes and cats? Before a theory is offered, I must first briefly examine the evolutionary history of these carnivores and second the taxonomy of Canidae. Most paleontologists agree that canid and felid carnivores evolved from a common ancestral group, the miacids, approximately 40 million years ago. These miacids were small, weasel or genetlike carnivores. They had an elongated body, a long tail, short limbs, and frequently semiretractile claws. They were probably forest dwellers that preyed on insects and other small animals.

The second point rests on the fact that I agree with several other specialists that the systematics of Canidae may need revision. Currently, the classification of wild dogs is based on S.G. Mivart's 1890 monograph, which examines anatomical differences and arbitrarily places a great deal of emphasis on differences in dentition. The use of other information—for example, biochemical similarities and behavioral criteria—to determine phylogenetic relationships in Canidae has received little attention. But recently Alfredo Langguth has taken new information into account and offers a classification of Canidae that is radically different from Mivart's. Although his work is still in progress. Langguth tentatively concludes that there are two main groups of canids: Group 1, the *Vulpes*-like foxes, which includes *Vulpes, Otocyon* (bat-eared fox), *Urocyon* (gray fox), *Fennecus* (fennec), and *Alopex* (arctic fox) and Group 2, the *Canis*-like canids, which includes *Canis, Cuon* (dhole), *Lycaon* (African hunting dog), most South American wild dogs, and others. The fact that Langguth identifies the *Vulpes* group as a distinct taxonomic subgroup of Canidae plays an important part in my hypothesis.

Using this information a theory can now be advanced to explain the behavioral and morphological similarities observed between foxes and cats. I suggest that foxes of the genus *Vulpes* differentiated early from the rest of Canidae and retained certain of their miacidlike characteristics—for example, a long tail, small foot pads, semiretractile claws, and long vibrissae. While the rest of the Canidae family evolved differently, the foxes went on to evolve catlike hunting equipment in their morphologies and feline hunting strategies in their behavioral repertoires. This convergent evolution is expressed to varying degrees among fox species and may be most strongly expressed in the red fox.

What ecological selection pressures would cause foxes to evolve in this way? As many food-habit studies show, foxes in general and the red fox in particular hunt many of the same kinds of prey as small cats. To escape both fox and cat predation, their prey probably use similar antipredator devices. To capture their prey, foxes and cats have evolved a number of similar features. Both have the interrelated features of a vertical-split pupil combined with a well-developed tapetum lucidum for good nocturnal vision; long vibrissae; long, thin canines and the lack of head-shaking for an effective killing bite; a fine sense of balance; soft, hair-covered feet; and quick reactions—all of which facilitate stalking and surprise attack.

Consequently the red fox and perhaps other fox species are as

Some of the fox's social displays also appear catlike. In the lateral threat, the dominant fox stands broadside to its opponent, arches its back, curves its tail, and bristles its fur.

much exceptions to the canid taxonomic lineage as the cheetah and lion are to the felid lineage. The red fox is a member of the Canidae, but because it hunts the same prey as small cats the red fox has evolved to become a very catlike canid.

5 *Evolving A Better Mousetrap*

In some vague, inexplicable way it reassures me to know that it is not unique to human nature—this arms race business. Whenever a field biologist takes a careful look at nature, he seems to find it: one group of living creatures inventing new lines of attack or defense, another group coming up with ingenious ways to circumvent the other's strategy. Whenever you find two groups competing—whenever there is an eater and an eaten—there will be evidence of an arms race.

Consider grasses—yes, lowly grasses. Even as these plants evolved, a number of animals discovered they were highly nutritious food. Grasses were being munched to death by an ever-increasing hoard of mouths nibbling their way across the prairie. The plants' response was insidiously clever: they evolved microscopic, rock-hard particles embedded along the length of their stems. Besides offering grasses partial protection from desiccation and abrasion, these silicon deposits also tended to wear down and fracture the teeth of animals that fed on the plants. Consequently, over time many animals evolved an avoidance of grasses. This antigrazer strategy seemed to work until certain animals—first the bovid family (sheep, antelope, buffalo), then horses and elephants—independently came up with a counterstrategy that was equally good. Gradually each evolved oversized, high-crown molars with specialized grinding ridges reinforced with a highly resistant enamel to contend with the crystalline fragments in grasses.

But the battle continued. Some grasses evolved poisons to weaken or kill grazers; the animals then learned to avoid these poison-laced plants. The ruse worked; for awhile the grasses seemed to have won. Then by chance a herbivore was born with a natural antidote to the toxin. This animal happened to carry a proteinaceous enzyme in its blood, a mutant molecule that could break down the belligerent poison and render it harmless. Now this grazer and all its descendents could eat the untouchable plant with impunity.

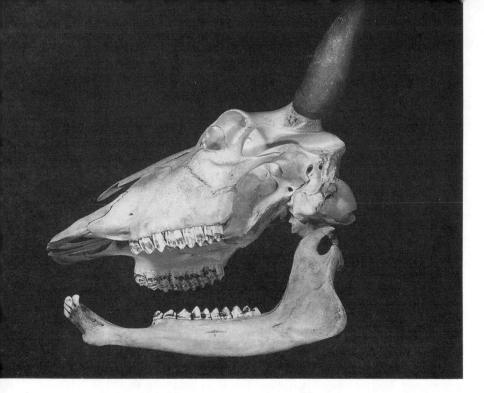

Like many other ungulates, bison evolved high-crowned molars to accommodate the wear caused by crystalline fragments found in grasses.

Point. Counterpoint. The strategy is as old as the contest of life against life. As soon as one group of living organisms begins to exploit another, the evolutionary arms race commences.

Consider bees. They evolved their venomous sting to defend their communal hive with its reservoir of honey. The bees' swarming attack is effective against most enemies—so effective that the insects flaunt yellow and black uniforms to warn opponents of their presence. Most animals respond by fleeing, yet bears evolved an immunity to the sting. To the bear a bee sting is less than a pin prick; the bear continues to eat, consuming honey, wax, and inhabitants of the hive with no apparent sign of discomfort. Will the bees by chance someday evolve a more effect venom—a poison that can stop even bears? Will the bears counter with an enzyme that can neutralize even this supervenom?

This warfare is as old as the tournament of life. The theory of an arms race among animals, however, is not universally accepted among biologists. Some question whether indisputable evidence for it can be found in the fossil record; other scientists contend that the theory can explain much, especially about predators and prey. The latter group maintains that in their evolutionary contest both contestants are constantly evolving new adaptations to thwart the interests of the other side. The prey evolves new and improved defenses that work temporarily until the predator in turn acquires techniques to overcome these defenses. The contest continues until one of the opponents evolves a "trump card"—an adaptation that the other side cannot match. If the predator comes up with a fail-safe way of hunting its prey, it may hunt the quarry to the point of extinction. Ironically if this happens predator and prey may both become extinct.

Man in his arms race also seems to be flirting with extinction. The uniqueness of man is not that he could destroy himself as a result of his military strategies, or even that he is waging a war against himself rather than another species. Intraspecific warfare can be observed in many species. The lamentable uniqueness of man is that his nuclear weaponry has reached such a stage of sophistication that if deployed it would carry into extinction not only man but many other species on earth.

Fortunately among animals these conflicts often have a more productive outcome. If by chance the contest stays in balance, if every adaptation of the predator is matched by an equally good counteradaptation of the prey, then over eons of time this competition will lead to the advancement of both predator and prey. This process has developed graceful, skilled predators and equally graceful, alert prey. It gives us the speed and coordination of the cheetah as well as the high-spirited alertness of the Thompson's gazelle. It evolves comraderie among wolves as well as the phalanxlike formation that muskoxen use against the wolves' attack.

Now let us consider what may seem to be a humbler contest—the tournament between foxes and mice. In fact, this confrontation is not as humdrum as it may seem, for ordinary field mice are challenging opponents. These tiny creatures are hunted by almost everything—hawks, owls, cats, weasels, wolverines, as well as foxes—and have shown amazing ingenuity in coping with this predatory peril.

Among their defense tactics are camouflage coloration, bulging eyes for wide-angle vision, an acute sense of hearing, split-second escape reactions, the use of burrows, and a detailed knowledge of their home range for rapid escape.

This last strategem needs closer examination. Lee Metzgar, a researcher who has spent years learning to think like a mouse, has shown that mice on familiar terrain are less vulnerable to predators than those on unfamiliar ground. Metzgar demonstrated this point in a simple yet elegant experiment. He built an indoor arena so that he could control light, wind, and temperature and planted it with local vegetation to make it as natural as possible. Then he released a group of white-footed mice and gave them several days to explore the area. A second group of mice was given only minutes to learn their way around; then he released a screech owl into the room. Metzgar found that on the average the predator caught five "inexperienced" mice for every "experienced" mouse captured. The results suggest that the "experienced" animals knew the terrain, knew where their escape burrows were located, and could escape more effectively than the "inexperienced" ones who became owl fodder.

Bill Bradley, pro basketball player turned senator from New Jersey, said that his success on the courts was due to playing with "a sense of where you are." Bradley believes that if you are going to score, you must always have a sense of where the hoop is and the position of the other players. I hope Senator Bradley will not be offended if I say that this is a very mouselike strategy. Having a sense of where you are on your home court is the tactic by which mice live or die.

A mouse's ability to learn and remember terrain is impressive. J.L. Kavanau, another researcher, built a mouse supermaze. It was meant to challenge the mental capacities of any animal, with 427 meters of passageways, 1,205 90-degree turns, and 445 blind alleys. Yet Kavanau found that common field mice, receiving neither reward nor punishment, learned to run this labyrinth forward and backward within two or three days. Mice enjoy learning terrain; their ability to create a mental map of their home range and to remember every nook and cranny is a topographic faculty that the human mind does not seem to possess. It is a skill that mice have evolved over eons of time to keep themselves out of the jaws and talons of predators.

Even if a mouse is captured by a fox, the mouse still has two defensive ploys: to fight back or to remain immobile. When a mouse fights back, I have observed that a young, inexperienced fox on occasion abandons the prey or hesitates long enough so that the mouse can escape. The other tactic involves the immobility response. Many animals, once caught, seem to enter a deep sleep so that they appear dead. Predators sometimes relax at this point, put down the prey, and investigate their surroundings. An unwatchful fox may then give the mouse time enough to regain consciousness and escape.

If the eleventh hour reprieve fails, there is yet another important tool in the mouse's arsenal. This tool does not reduce the chances of being caught but lessens the implications of being eaten. It is called "breeding like crazy"—or more formally, compensatory reproduction. Animals that are heavily hunted, such as

Small rodents, such as this deer mouse, exhibit a number of antipredator devices, including camouflage, wide-angle vision, acute hearing, and split-second flight reactions.

mice, rabbits, and other small game, compensate for predation by "over-reproducing." They produce a surplus of young—many of whom are inevitably eaten—with the net result that a sufficient number still survive to become adults the following year.

Armed with these ploys, mouse sets out to meet fox. How does fox respond? What counterstrategies has the red fox evolved to overcome these microtine defenses?

On my desk scattered in front of me are 434 large-format index cards. On each is typed a description of a fox making a hunt. Some of these hunts are successful, others not. Some hunts are after mice, others are directed at shrews, songbirds, insects, snowshoe hares, or red squirrels. Some lasted only a few seconds—for example, a fox biting frantically at a sparrow that flew out of a bush. One hunt lasted more than an hour—the fox patiently lying in ambush behind a shrub waiting for an unsuspecting squirrel to come near. Each card contains a detailed description, transcribed from a recording made in the field, of a fox performing an individual hunt. In total, 434 observations slowly compiled over 14 years—a stack of data now dog-eared and soiled from being repeatedly studied. The edges of these cards, however, are purposely ragged and serrated. These perforations allow me to sort the hunts quickly and effortlessly according to any feature I want.

I stick a long needle through the perforation designating mouse hunts, pick up the entire pile, and shake—83 cards drop onto my desk. Now using this stack, I select the hole for successful hunts, insert the needle, shake, and 23 cards fall out. I pick one up and look at it. It is a hunt by The Prince, a young male fox with a particularly arrogant manner. I sit back in my chair and remember golden, sun-drenched autumn days I spent following The Prince, watching him become a proficient hunter. I recall the exhausting physical exertion it took to keep up with this energetic slayer of small game and all the things that I observed and learned that autumn. I remember particularly The Prince's idiosyncrasy of hunting in the same location day after day until he had captured the rodent living there, almost as if it were a matter of honor. No other fox I have followed has shown such dogged determination. The card reads:

> 30 October 1974. 9:15 a.m. There is a fresh skiff of snow on the ground, and The Prince is hunting more seriously now than before the snow came. I've seen in him few signs of playing today.
>
> He walks along a hiking trail in the woods until he hears something off to the side and his ears go very alert. Turning perpendicular to the path, his head goes into mousing position and he listens intently, trying to locate the source of the sound. He cocks his head one way and then the other, and his rump sinks down as he slowly brings his hind feet up close to his forefeet. Now he stares intently at a spot and flicks his tail back and forth excitedly, very catlike. He pumps his front legs up and down slightly, nervously, crouches deeply, and then lunges about three yards in front of him and bites down into the grass. Holding his bite for several seconds, he then brings his head up high and looks about, having missed the prey. Suddenly he makes a second lunge to the other side of the grass tussock, bites down, and then circles around the grassy mound. He knows the prey is in there somewhere . . . he circles the mound again and then stands for 45 seconds, sniffing first with his head low and then high. Suddenly he lunges into the

middle of grasses and pushes his nose
through them. Then quickly he marches
around the mound, trying to cover all sides
at once . . . he waits again and then quietly
stalks to the other side of the mound. All of
a sudden the mouse makes a run for it; and,
seeing it, The Prince lunges long and hard
and bites down on the mouse while both are
running. Biting the prey hard several times,
he then tosses it up into the air and catches
it with his mouth; then he carries the prey
up onto the trail, sets it down, and watches
it. The prey is a short-tailed vole, perhaps
Microtus. . . . The hunt was a long one, last-
ing 2 minutes and 43 seconds.

The prey does not move. The Prince noses
it; the prey still does not move. Now picking it up in his mouth, he smells where it
lay on the ground and urine marks the spot.
He then carries the prey 75 yards down the
trail and 10 yards off to one side and carefully caches it there. After that The Prince
goes right on hunting.

In this hunt all the characteristic features of a small-burrowing-
mammal hunt are present: the fox hunts from a trail, searches for
the prey with its head in mousing position, lunges, and tries to
pin the prey to the ground with its forepaws. Combined, these
features function to minimize the amount of noise that the fox
makes; small burrowing mammals are sensitive to noise, it is the
major way they detect predators before they themselves are detected. If the fox snaps a dry twig or breaks a brittle leaf, it can
send the mouse fleeing underground. But by lunging the fox soars
over sticks, leaves, and small shrubs on its deadly flight toward
the prey. Lunging allows the fox to maintain its surprise attack to
the very end. In densely vegetated environments, such as the
boreal forest, lunging seems to be a consistent component of
small-burrowing-mammal hunts: I saw it in more than 90 percent
of these hunts but in only 13 percent of the hunts after other prey.

Let me try to convey what the red fox does when it makes one
of these long, graceful lunges after a mouse. Imagine yourself at a
party at a friend's home. During the latter part of the evening,
you find yourself standing in stocking feet at one end of his living
room. Your friend is at the other end, placing a spoon on the
carpet. You have wagered that from a standing start you can leap

across the room and pin that spoon under your toes. Between you and the other side of the room are several easy chairs and your friend's color television set. You are going to leap over these things and pin the spoon to the carpet. Having bet money, you cannot afford to miss; the room is quiet, everybody awaits your try. Oh yes, to make things more realistic you cannot see the spoon but have to locate it by sound. Your friend has tied a string to the spoon and is slowly dragging it across the carpet. Are you ready?

This is essentially what a red fox does when it makes one of its long hunting lunges. How is it possible? How, from a standing start, does a fox lunge 15 feet across level ground, or 25 feet downhill, and pin a mouse under its front paws? After seeing several of these lunges, I became intrigued by this question and set out to understand how the fox is capable of performing such a feat.

The Fox as Guided Missile

It is night again. Late. Bitter cold and yet silent outside. I sit at my desk alone; everyone else in the cabin sleeps peacefully. I can hear the jackpine logs in our airtight wood stove crackle, break apart, and settle against the hissing aspen logs.

On my desk are a physics book, digital calculator, and graph paper. The same stack of tattered index cards sits in the corner, and several of them are spread out before me. The book is opened to a chapter on guided missiles. Strange bedfellows—the arctic cold; the quiet of a boreal night; yellow, aging index cards; a digital calculator; and a 20-year-old physics book—and yet it is the strange relation of these items that I am exploring.

The quiet winter woods are a fine place to think; there are so few disturbances. Earlier that evening while snowshoeing it had occurred to me that when the fox leaves the ground in one of its long, graceful lunges it acts like a guided missile aimed at a small target. If I can understand the forces that determine the flight of a projectile—be it rocket, bullet, or lunging fox—then perhaps I can understand how red foxes, unlike any other canid that I know of, can soar these distances.

The physics book is open to an ominous-looking formula. It states:

$$R = \frac{(F \cdot T)^2}{g \cdot m^2} \text{ sine } 2\theta$$

The equation looks complicated; part of me wants to forget the book and go out and watch the moonlight dance off the snow. Yet I press on.

To capture small rodents or shrews, a fox normally crouches deeply, lunges, and with a stab of its forepaws pins the prey to the ground.

With cold logic the book tells me that in the formula R equals the range of the guided missile, or how far it will fly. F represents the force of its engines; T is the amount of time the engines are firing; g is the pull of gravity on the missile, bringing it back to earth; and m is its weight (mass), fuel and all. The complicated-looking thing over on the right-hand side, that *sine 2θ* business, simply refers to the angle at which the missile takes off. I reflect on all this, think I understand, and wonder what comes next.

Now the book tells me that if you fired the projectile straight up in the air (takeoff angle equals 90 degrees), it will go straight up and come straight down. In relation to the surface of the earth, it will not have traveled anywhere; its range will be zero. If you fire it parallel to the ground (takeoff angle zero degrees), soon the pull of gravity will force it into the ground and it will not get very far either. The book says the range of the missile will be greatest if it takes off at an angle of 45 degrees. I can accept that; it seems reasonable. I wonder if the fox when it makes one of its long lunges takes off at an angle of 45 degrees.

I get up from my desk, put more wood in the airtight, and make myself a cup of tea. I sit down at the table and write the equation out on a piece of paper, sip my tea, and look at the paper. Slowly I come to realize that the formula says something very simple about the fox. I imagine a fox lunging across level ground on a calm day so that wind and the slope of a hill are not factors. Assuming these conditions I realize the formula states that only five factors can affect the range of a projectile, be it flying fox or cannonball: the pull of gravity, the mass of the projectile, the amount of force the projectile exerts against the ground, the amount of time this force is exerted, and the projectile's takeoff angle. Because the fox can do nothing about the force of gravity, it is left with four strategies to maximize the length of its hunting lunges:

1 The fox can take off at an angle of 45 degrees.
2 The muscular force that the fox exerts against the ground before taking flight could be made as great as possible.
3 The amount of time that the fox exerts this force against the ground could be made as long as possible.
4 The fox could lessen its weight. This tactic would be effective only if the fox could reduce its weight without lessening the power of its lunging muscles.

During its evolution the red fox could work with these four strategies, but nothing else would matter because nothing else influences the range of a projectile. The formula did not show me what adaptations the fox has evolved for lunging, but it did give some guidance about where to look.

Indeed this formula was to direct my efforts for several years. It would lead me into making detailed measurements of lunging foxes. It would take me into the dusty vaults of museums to compare the skeletons of red foxes with those of other canids—all in an effort to understand how the fox soars silently aloft, circumventing many of the mouse's antipredator devices, to land and pin its small prey to the ground.

I began my inquiry by measuring the fox's takeoff angle. I had several movie sequences of foxes making long lunges. This footage is extremely difficult to get. It requires countless hours of following a fox, patiently waiting for this fur-covered catapult to launch itself. Over the years, however, I had been able to capture seven of these long lunges on film, and I felt these sequences would allow me to measure accurately the fox's propellant angle.

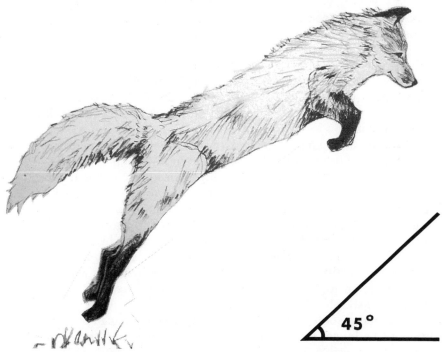

To maximize the range of its hunting lunge, the fox should lift-off at an angle of 45 degrees.

To do this I carried my valuable films to a frame-by-frame analyzer. This machine lets a person look at each frame of a movie and trace a drawing of it on a small screen. I searched my first film strip until I found the frame where the hind feet of the lunging fox were just leaving the ground—the moment of lift-off. I could use this frame to measure the fox's launching angle. But first I had to draw on the screen a line that represented the true horizontal, that is, the line exactly perpendicular to the pull of gravity on the fox. Without this line I could not proceed, yet I had no precise way of drawing it.

After several hours of thinking about this problem, I came upon a solution. I remembered that trees that are healthy and uninjured tend to grow parallel to the force of gravity. Soon I was back at the analyzer with the frame of the launching fox focused on the screen. I took all the healthy, straight trees in the picture and drew a line perpendicular to their trunks. Averaging these lines gave me a good estimate of the true horizontal. Next I drew a line parallel with the back of the hunting fox. Where these two lines met was the takeoff angle of the lunging fox. I simply measured how large this angle was with a protractor. I repeated this process on all seven movie strips. When I was finished, I found that the fox's average takeoff angle was 40 degrees ± 6 degrees. The fox is indeed a good marksman.

On the average the fox was slightly below the optimal value of 45 degrees. There may be a good reason for this. The higher the fox goes during a lunge, the more it can be blown around by the wind, and the more likely it is to be seen by the prey. I would expect the fox to use the lowest takeoff angle it could and still reach the prey.

Certainly a fox does not calculate its angle before launching, but it appears to be somewhat sensitive to its lift-off angle and the way it affects the range of its lunge. For example, when a fox makes a short pounce, leaping only one or two yards, the analysis of films shows that its angle may be as low as 10 to 15 degrees. On the other hand, when the fox is breaking through snow with an icy crust and needs maximum downward thrust, he may use an angle as high as 80 degrees so that he lunges almost straight up and down. But when lunging for maximum distance, the fox takes off at an angle much closer to 45 degrees—just as a good guided missile should.

What about some of the other strategies? The fox can lengthen its hunting lunges by (1) making the force it exerts on the ground as great as possible or (2) extending the amount of time it exerts this force. I decided to look at these strategies together. To understand them better, I examined some animals specialized for hop-

ping—for example, the kangaroo, frog, and rabbit. These saltatorial animals show some interesting features in common. All have
massively developed hind legs, and all sit in a deeply crouched
posture. Clearly these animals have maximized the *force* they
exert on the ground by evolving massive leg muscles, but how do
they extend the *time* over which this force is exerted? Imagine
painting a small red dot on the rump of a frog. Where this dot is
when the amphibian is sitting and where it is just before the
animal becomes airborne measures the amount of time the frog
exerts force against the ground. The frog can extend this amount
of time either by starting its hop from a more deeply crouched
posture or by making its hind legs longer. Obviously during frog
evolution both characteristics were developed to an impressive
degrees.

How does all this apply to the red fox; how froglike is it? The
fox does not seem to have the massively developed hind legs of a
saltatorial animal. It relies too much upon running to have
evolved the legs of a kangaroo. But perhaps the fox has other, less
obvious muscular features that lengthen its lunges. The fact is I
don't know. To answer this question conclusively would require a
comparative morphological study of canid muscle systems, and
this research has not been done. Until it is carried out (and my
hunch is that it would show some interesting adaptations), we do
not know if the fox has evolved any special muscular features for
increasing the force of its lunging muscles.

The only question I was able to look at is: Does the fox show
adaptations for extending the amount of time it exerts force
against the ground before becoming airborne? While browsing in
the claustrophobic stacks of a university library, I quite by acci-
dent came upon a reference to Milton Hildebrand's 1952 analysis
of the body proportions of wild canids. It was an old study but
sounded interesting. Returning to the sunlight for the proper call
number, I then descended and, like one of Kavanau's mice, navi-
gated through the iron maze until I reached the proper blind alley
and plucked a dark green volume off the shelf.

Hildebrand's study was fascinating; it made me forget the
dusty, slightly emetic atmosphere of my book-filled labyrinth.
Hildebrand had made his analysis on museum skeletons during a
time when field observations on these wild canids were meager
indeed. Regarding the red fox, he concluded that it was the only
canid species in his study to show relatively long hind legs. He
stated, "In evolving longer hind limbs this fox has departed from
a body proportion common to several closely related canid species." Hildebrand went on to say that he was unable to offer any
explanation for this feature in the red fox. That to me was the

intriguing part; Hildebrand had inadvertently discovered an important adaptation in the fox but was not able to explain its significance. For an experienced fox follower like myself, it was not difficult to suggest a reason why this characteristic had evolved: foxes lunge from a deeply crouched posture; this characteristic combined with their long hind legs increases the amount of time they exert force against the ground and thus lengthens the range of their hunting lunges. Under all that luxuriant fur, the red fox does seem to hide a froglike characteristic or two.

The fox is now left with one strategy: if it can reduce its body weight without a loss of muscular power, these features would also increase its pouncing range. Is there evidence to suggest that the fox had a lighter body weight than one might expect? I did not have to follow a fox for long before becoming impressed with its legerity; however, because this might be the prejudice of an addicted fox follower I searched for more objective evidence. This effort led me to look at the foxes of Europe.

European red foxes are the same species as North American red foxes but are slightly bigger. When I compared European red foxes to a small race of coyotes that inhabit the dry, arid lands of the American Southwest, I found that the male red foxes are approximately the same size (same body length) as the female coyotes. But how do they compare in weight? The foxes weigh only half as much as these coyotes. Other comparisons led in the same direction. Several breeds of dogs—for example, the Irish terrier and English foxhound—have similar linear dimensions to red foxes. Yet the suggested show weight for these dogs (22 to 28 pounds) is nearly twice the average body weight of red foxes (12 to 15 pounds). This evidence suggests that foxes may be relatively light for their size.

If the red fox is disproportionately light compared to some other canids, what mechanisms during evolution could the fox use to lighten its body weight without losing muscular power? What cargo could the fox afford to jettison to become the featherweight of wild canids?

An insight came to me one day as I watched Rose, an older female fox, scavenge meat from a deer carcass. Wolves had killed this deer quite close to some cabins and therefore had not cleaned it up to their usual extent. When Rose found it early on a brisk winter morning, a fair portion of meat was left on the bones. The fox was definitely hungry; she had been hunting vigorously just before discovering the wolf kill. Yet despite her hunger Rose was not able to consume much more than 1 to 1¼lbs. of meat. When she had eaten her pound of flesh, she was sated. However, rather than surrender the carcass to the marauding ravens she began to cache little bundles of meat throughout the surrounding woods, setting up provisions for later use.

Rose's dainty appetite is quite a contrast to the etiquette of wolves. A wolf commonly eats 18 to 25 pounds of meat, or approximately 20 percent of its body weight at one sitting. The largest amount I have ever seen a fox eat at one time is 1½lbs. of meat, or approximately 10 percent of its body weight. This difference is due to the fact that the fox's stomach is relatively much smaller than the wolf's. I believe that red foxes have evolved a small, light stomach as a means of reducing their body weight. The fox's stomach also correlates well with the "bite size" prey it usually catches. In essence, red foxes are only "nibblers" when compared to wolves.

The fox has also lessened its body weight by reducing the mass of its skeleton. Milton Hildebrand in a second study also shed considerable light on this point. In 1954 he analyzed the relative width of bones from skeletons of numerous canid species. Hildebrand did his study by choosing a bone—for example, the femur— and then drawing this bone from the different canid species to exactly the same length. Thus in his drawings the femur from a wolf was the same length as the femur from an arctic fox. Having done this he could then compare the relative widths of these bones. The pattern that Hildebrand expected to find was that the massiveness of these bones would increase to support the proportionately greater body weights of the larger animals. In fact, he did observe such a relationship in the spinal columns of these species: the vertebrae became relatively more massive as the body weights of the animals increased. On the other hand, the limb bones showed more variation, and massiveness in these bones was not always a function of body size. For example, the two species that showed the proportionately narrowest limb bones were the shortest and the tallest canid species in the study, namely, the North African fennec (*Fennecus zerda*) and the South American maned wolf (*Chrysocyon brachyurus*). In addition, the bush dog (*Speothos venaticus*) from South America had relatively the widest limb bones even though it was one of the smallest canid species represented. Regarding foxes Hildebrand showed that, after *Fennecus* and *Chrysocyon*, foxes of the genus *Vulpes* exhibited some of the most slender limb bones of any canid species examined. Specifically he found that a number of bones, including the sacrum (hip bone), sternum (breast bone), scapula (shoulder bone), lumber vertebrae (lower back), baculum (penis bone), and ulna and fibula (lower leg bones), are significantly reduced in their relative width as compared to those of other canid species. In other words certain bones in the fox's skeleton

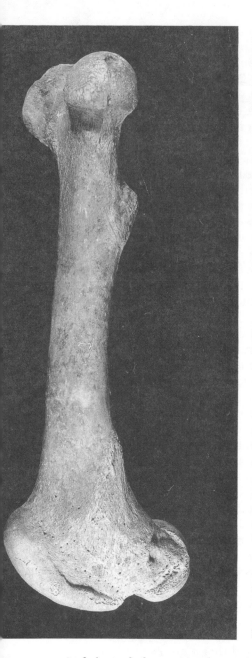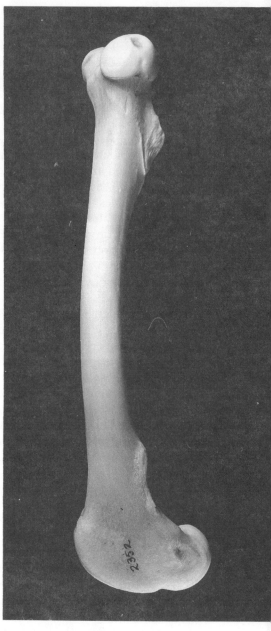

Inch for inch, large animals weigh proportionately more than small animals. The femurs from a bison and a deer shown above were photographed so that they appear to be the same length. However, the bison femur is much thicker and heavier to support the proportionately greater body weight of the bison.

are surprisingly slender, a feature that would contribute to a reduction in the body weight of the red fox.

Having learned this from Hildebrand's study, I began to wonder if red foxes would show any birdlike characteristics in their skeletons. Birds are well known for their extremely light yet strong bones—an important adaptation for flight. They achieved this skeletal feature by evolving a sophisticated buttressing system inside their bones so that heavy calcified material could be lost without a subsequent loss in the functional strength of the bone. These lightweight characteristics of a bird skeleton are described as osteopenic properties—a word meaning, literally, the reduction of bony material.

Years earlier I had collected and preserved the complete skeletons of a male coyote and a male red fox. I knew these skeletons had been prepared by the same process and degreased to the same extent so that the weights of these bones could be compared. I did not include the ulna and fibula in my study because Hildebrand had shown these bones to be disproportionately slender in the fox; I did, however, compare the four other limb bones to see if the fox's long bones are lighter than we would expect for a mammal of its size.

To carry out my study, I had to use Alfred Taylor's "rat to cow" curve. In 1935 Taylor, a morphologist, did a classic study on various-sized mammals. Others had established that during evolution as animals increase in size their body weight multiplies approximately as the cube of any changes in their linear dimensions. This relationship is the result of simple physics—for example, a wooden block that doubles its length, width, and height will contain eight times as much volume, and all things being equal will weigh eight times as much. Based on this principle, Taylor reasoned that the skeletons of big animals have to be proportionately much larger and heavier than that of small animals in order to support their greater mass. He expressed the results of his experiment this way: the limb bones from a rat show 10.6 cm^2 of surface area for every gram of bone while the limb bones of a cow show only 0.69 cm^2 per gram of bone. Taylor studied mammals of different sizes, and from these results he produced his noteworthy curve. For example, he studied different kinds of domestic dogs and came up with an average value of 2.96 cm^2 per gram of bone. Taylor never studied the skeleton of a red fox, but his curve predicts that the limb bones from a mammal of this size should show 4.91 cm^2 of surface area per gram of bone. What does the fox actually show?

Now here is the rub! To obtain the value for the fox, it is necessary to determine the surface area of its limb bones. The "easiest" way to do this is to meticulously cover the bone with adhesive tape, making sure that every nook and cranny and every complex bump of every ridge is carefully covered in tape—and there must be no overlaps. Having done this you must with equal care determine the exact weight of 1 cm² of tape. Then after all that work, you must strip all the tape off the bone, weigh it, and calculate the exact surface area of the bone. Finally you must weigh the bone and divide this weight into the surface area. This will give you Taylor's value for the bone from the fox.

Taylor's method was corporal punishment. I spent a number of bleak winter evenings with my coyote and fox bones spread over the kitchen table. Razor blades and drafting knives helped me painstakingly cover each bone with its antiseptic white coat. My wife laughed as I cursed some minuscule recess of bone, trying to get it precisely covered with tape; however I endured and my efforts led to the following results.

Bone	Weight (gm)	Surface Area (cm²)	Index Value (cm²/gm)
Red Fox			
Humerus	10.3	64.0	6.21
Femur	10.5	70.4	6.71
Radius	5.6	36.9	6.59
Tibia	10.9	67.4	6.18
Average			6.42 ± 0.27
Coyote			
Humerus	38.1	138.8	3.64
Femur	40.1	156.1	3.89
Radius	20.9	82.4	3.94
Tibia	36.7	121.0	3.29
Average			3.69 ± 0.29

Basically I found that the coyote's limb bones gave an average of 3.69 ± 0.29 cm² of surface area per gram of bone. This figure is close to Taylor's predicted value of 3.88 cm² per gram of bone for an animal the size and weight of a coyote. By contrast the limb bones from the red fox skeleton gave an average of 6.42 ± 0.27 cm² per gram of bone. This value is considerably higher than Taylor's predicted value. Consequently the fox shows over 70 percent more surface area per gram of limb bone as compared to the coyote or domestic dog, and 30 percent more than expected for an animal the size of a fox. While this is only a preliminary test of the hypothesis and more data is needed, it does suggest that the red fox's limb bones are lighter than we would expect for a mammal of its size.

My physics formula had shown me that the fox had just four strategies by which it could lengthen its hunting lunges; I found evidence that it had perfected three. A feeling for the oblique, a deeply crouched posture combined with long hind limbs, a "mouse-size" stomach, and a slender, light skeleton were attributes that natural selection had developed in the fox to allow it to perform long, graceful hunting lunges. Some of these evolutionarily sculptured features are undoubtedly useful in other contexts as well. For instance, the fox's dart-about running, either when chasing rabbits or eluding coyotes, probably makes use of its light body weight and long hind legs. But all these features definitely come together to help the fox soar quietly aloft, float silently past the antipredator defenses of small mammals, and pin its unsuspecting prey securely to the ground.

But make no mistake. The contest between mouse and fox remains balanced and challenging for both. Let the fox tread on a dry leaf or crack a brittle pine needle, and the mouse can disappear safely through one of the semipermeable pores of its home range; let it miss the prey by a fraction of an inch, and the lightening-quick reactions of the mouse will deliver it safely underground. Fox and mouse are still engaged in a balanced evolutionary contest, a tournament worthy of the fettle of both.

Why does the fox lunge? To catch mice, insectivores, and other small burrowing mammals of course. After all the red fox is the ultimate canid mousetrap. However, there is more to it than that. Foxes probably also lunge for the thrill of it. They soar, tracing their volant paths because they find flight exciting. The urge to lunge appears early in the life of the fox. Frequently I have watched eight-week-old fox pups leap down the den slope, collide with a littermate, and try to pin their sibling to the ground. The fox cub pounces—perhaps with little understanding of what it is doing or why it enjoys doing it. Watching these young foxes perform their atavistic lunges, I became convinced that the urge to lunge was born long ago in the fox; it is an action embedded deep within the fox's muscles and bones, a penchant seated deep within the fox's psyche. Foxes lunge for the same reason that mosquitoes bite. It is an action that they have evolved to do, a behavior by which young foxes will in a short time provide food both for themselves and for their own growing families. It is an act, like many others, of an individual animal or family competing for needed resources.

When I watch an ember-colored fox launch itself in a graceful arc, I feel that same electrifying sensation I feel when I watch mountain sheep ram heads or a pack of wolves howl or people philosophize. Why do foxes lunge? Why do rams break skulls against each other and wolves harmonize in syncopated rhythms and men argue late into the night? Because these actions were meant to be—they are the particular excellence of the species, the pursuit each of us is meant to follow.

Holding food in its mouth, the fox digs a hole to make a cache. The fox digs gently, keeping the excavated dirt close to the edge of the hole.

6   *Cutting Your Losses*

It is October. I am standing in an aspen woods, ankle deep in gold leaves, surrounded by the ruins of summer. I am watching a buff-red fox I named Amber cache a mouse she has just caught. She circles around among the trees, checking out places and deciding where she is going to put her prize. I have seen foxes cache their prey before hundreds of times; but I am curious to know how this six-month-old vixen—a fox who has never seen snow or ice and who has been catching her own prey for only a short time—will cache. The fox noses around shopping for her spot, and I laugh softly as I think how I had judged caching to be a simple and straightforward behavior. That was in the beginning. Now I view it as a mysterious and complex vulpine ritual. I watch closely as the vixen selects a spot at the base of a large aspen and begins to dig.

Amber seems to cache almost by impulse, yet she caches correctly. She shows all the actions, if not some of the finer details, of a mature fox. She begins by carrying her prey away from where she caught it into a new stand of aspen. Now, still holding the prey gently in her mouth, she digs a mouse-size hole, but one deep enough so that when she is finished the prey will be covered by three to four inches of dirt. She does not spray the dirt back as she digs but piles it neatly next to the hole. Then placing her prey in its shallow grave the fox presses it down firmly into place; next she pushes dirt over the carcass and packs the soil down, stabbing at it with her snout. She uses her nose, an organ of extreme sensitivity, much as I would use a blunt stick to pack dirt into the hole. Finally, again with her snout, she rakes leaves and twigs on top of her cache and with various pokes and jabs rearranges this forest litter until her cache, at least to my eyes, blends perfectly with the surrounding forest floor. She makes a few final adjustments, lifts her head, looks down at her camouflage, and walks away, confident that she has at least one mouse securely tucked away for future use.

It took Amber only 90 seconds to complete these actions, which is about average, although I have watched foxes with

cashing

95

highly valued food items spend up to 15 minutes fussing over their caches. They do not bury the food any deeper, but they spend more time packing the dirt down and arranging the camouflage. I remember one cold November afternoon in 1977 watching My Friend, who was by then one of the oldest and most experienced foxes on my study area, make an extraordinarily careful cache. He was stashing the hindquarters of a snowshoe hare in a frozen alder bog. Digging into the snow, he carefully folded his rabbit into the hole. Then with his muzzle he pushed snow in on top of the rabbit and packed it down firmly—building up five separate layers on top of the prey and packing each carefully. Then using his nose like a small paintbrush he whisked it across the surface of the snow, removing all signs that he had something buried there. With increasingly delicate strokes, he made the surface of the snow smoother and smoother. Then ceremoniously he began to step backwards, retracing his steps. With each step he lowered his nose and brushed in his paw prints leading up to the cache. Here was a master at the game, maximizing his chances

A fox makes a winter cache, gently brushing snow over the buried food to camouflage it.

that his rabbit would still be there when he returned for it on a stormy winter night.

Of course what foxes are doing when they cache is putting provisions away for future use. As I have already mentioned, foxes procure their food through both hunting and scavenging. Upon rising a fox usually sets off on one of these food-getting activities. If its luck is bad, it may switch from hunting to scavenging or vice versa. Any food that it obtains it normally eats immediately until its appetite is sated, but after this point it often continues to hunt and scavenge. It caches any food it obtains during this time.

During lean and hungry times—that is, if hunting turns bad, if a fox injures himself, or if heavy weather sets in—a fox relies on his caches. I have found that a fox can maintain itself on approximately one pound of meat per day, and I have traveled with foxes on days when most of their daily rations have come from their buried stashes. The foxes' caches are of genuine value to them.

It helps to watch foxes cache 500 times or so, as I have done, because by then a person may actually begin to see this important vulpine operation for what it is and to understand the forces impinging on it. Niko Tinbergen calls this "exploratory watching". Tinbergen, one of the founders of ethology, says a scientist should watch a behavior over and over again until even the idiosyncrasies of the behavior make good, practical sense. A biologist should watch and think about an action until he or she understands why evolution has shaped and fashioned the behavior as it has. For just as evolution sculptures and perfects an animal's teeth to suit its diet, so does evolution sculpture the form of a behavior until it achieves its function easily and effectively. By watching a behavior over and over, one begins to understand why the behavior has evolved the appearance and shape it has. Niko Tinbergen considers this type of animal-watching one of the most difficult and important tasks that a naturalist undertakes when trying to understand how an animal and its world fit together.

The first few times I watched foxes cache I was intrigued by this clever maneuver, but I wondered, like many naturalists before me, if foxes ever really use these caches. How do they ever manage to relocate them? To foxes these questions would seem ridiculous. After following foxes for several months, I became impressed with a fox's ability to remember obscure nooks and crannies on the forest floor. In essence, foxes remember precisely where they put many of their caches and return to them within a few hours or a few days to consume the food. On the other hand, foxes sometimes leave their caches hidden for much longer periods of time, and consequently they may have difficulty remembering exactly where the caches are. Niko Tinbergen describes one such incident. One summer on the Ravenglass Dunes this great Dutch naturalist watched a red fox repeatedly stealing eggs from a breeding colony of herring gulls. The fox hid each egg at a new location, scattering its caches throughout the whole dune area. But it was not until two months later—when the gulls had left and the supply of fresh eggs and chicks had vanished—that the fox began to consume the buried eggs. Tracking this animal, Tinbergen saw that the fox now did not seem to remember the cache's exact location. Instead the fox trotted into the general area of the cache and then began to crisscross slowly back and forth with nose close to the sand until it smelled out the hidden egg.

Thus foxes seem to have several ways of relocating their caches. Primarily they remember precisely where they put them. If total recall fails, however, foxes remember the general area and locate the cache by smell. This back-up system may explain why foxes do not bury their caches any deeper than about four inches. This paw-depth burial represents a compromise in a difficult situation. On the one hand, the fox has to bury its food deep enough so that the many scavenging animals that rob the fox's caches are foiled. On the other hand, the fox does not want to bury it so deep that it cannot detect it with its nose if it forgets exactly where it put it. A blanket of dirt several inches deep seems to be the best compromise. It protects the cache from marauding animals and yet lets the fox locate the loot if it cannot remember exactly where the cache is.

The list of scavengers that rob a fox's cache is long, and in my area includes birds, such as ravens, crows, magpies, and Canada jays as well as mammals such as weasels, wolverines, coyotes, wolves, and bears. When I say that one of these animals "robs" a fox's caches, I want to make it clear that I do not mean to imply anything about the motives of the "thief" or even how the fox feels about this act. I am simply using terms like "rob" and "steal" as a shorthand way of saying that one animal takes a cache that it did not make. Pressure from these "robbers" makes the fox cache as carefully as it does.

There is one more scavenger that the fox must contend with—one that may be the biggest problem of all, namely, other foxes. In vulpine society there appears to be no hesitation about stealing another fox's caches. It seems to be done whenever one fox stumbles upon another fox's stash. The thieving fox may eat the cache or carry off the prize and hide it elsewhere for its own future use.

I remember one cache that changed hands several times. One day in late August, as the birch leaves were beginning to turn, I was traveling with My Friend when he led me into a dense spruce forest and up to an empty caching hole. At first he seemed agitated; but soon he began to smell the hole and the area around it until he picked up the spoor of the pilfering fox. Using his nose, My Friend began to track this robber, following its trail out of the spruce forest and through several hundred yards of muskeg. Finally he came to the place where his loot had been stashed. After excavating the cache, he sat down and ended the contest by consuming the meat. Then he rose, urine marked the caching hole, and walked off into the woods.

It was only after a number of months that I really began to look at this behavior and appreciate the more sophisticated aspects of caching. For example, I began to wonder why foxes always spread their caches out, putting each load of food in separate, well-spaced holes. What benefit do they gain from scattering? Another thing that intrigued me was the uncanny consistency with which foxes cache. Why do all foxes do it the same way? Certain actions always follow other actions, and I have never seen a fox vary this pattern. For example, a fox buries its food and packs the dirt down tightly. Only then does it begin to cover the food with leaves and litter. A fox never mixes these actions. I have never seen a fox push a mixture of leaves and dirt in on top of the food, give it a few pats with its muzzle, and leave it at that. I became curious to know why the fox's caching behavior follows such a stereotyped pattern. When a behavior is invariable like this, it does not necessarily mean that it is innate. It may or may not be. What it does mean is that there is some strong environmental pressure on the animal to make its behavior rigidly fixed; the animal cannot afford to vary this behavior.

Think of your own actions and the stereotyped quality of your morning ritual. Because a person rises, showers, eats the same breakfast, and glances at the same parts of a newspaper morning after morning does not mean that these behaviors are innately determined by the person's genes. There is simply a reason causing the uniformity of these actions. In this case it is pressure to perform these acts within a certain period of time and be on our way. It is the environmental constraint of a 7:57 commuter train that makes our morning actions stereotyped; not a question of these behaviors being genetically determined.

I suspect that the caching behavior of the red fox is passed from one generation to the next by a combination of instincts and learning. When I watch a fox pup barely six weeks old wobble out of the den and cache a small bone, it makes me suspicious that the red fox is somehow predisposed to cache. Somewhere in the brain of this young fox there is a set of connections that urges it to perform these actions. The young fox caches in an incomplete manner; it digs a shallow scrape and does not cover the bone completely with dirt. Yet these behaviors are unquestionably present in the kit, ready to mature, ready to be perfected by practice and learning. It reminds me of what Paul Leyhausen, the distinguished German ethologist, says about the nature/nurture debate. Together with ethologists P.P.G. Bateson, Jack Hailman, Peter Klopfer, and others, Leyhausen believes that there is an instinctual basis to almost any action that we or any animal learns. Rather than being based on a simple dicotomy (a question of instincts versus learning), learned behavior may always develop on an instinctual foundation. In fact, animals that learn the most may also inherit a more complex set of instinctual behaviors from their parents. According to Leyhausen instincts are often a template upon which learning takes place; they predispose the animal to learn certain important behaviors in the most adaptive manner.

Whether or not one accepts the theory that the fox's caching behavior develops from both an innate and learned component, my initial question remains unanswered: Why is the fox's caching behavior so stereotyped? The ultimate selection pressure must be some environmental constraint, and the most likely candidate is the continuous threat of theft. But I wished to understand this competition better; I wanted to understand what specific features in the fox's caching behavior have evolved as antirobbery devices and how each has worked. I began to feel like an engineer evaluating the design of a new product, testing each of its parts and then testing the whole package for soundness, originality, and efficiency. To do this I knew I somehow had to disassemble the caching behavior of the red fox and test each of its components in as natural an environment as possible.

Niko Tinbergen came to my assistance again. If he had not done his famous field experiments in which he painted hen's eggs to look like gull eggs so as to test the effectiveness of their camouflage, I may not have known how to proceed. If he had never made artificial bird nests to test the effectiveness of spacing as a means of avoiding predation, then perhaps I would never have known how to explore the design of the fox's caching behavior. But of course Tinbergen had done these elegant field studies, and I, like a whole generation of biology students, had read and been impressed with the clarity and precision of his work.

So I set off in the summer of my fox-watching days into a remote area of Prince Albert National Park, determined to probe the design of the fox's caching behavior. I went to an area that few people visit. The ones who come travel by canoe. They come and enjoy the wilderness, to listen to wolves howl, to watch moose feed in their summer ponds, and to admire the black claw

Pups as young as six weeks of age cache inedible remnants of food found around the den.

scars left behind by bears in the chalky white bark of aspen. I came for all these reasons, but I also came to imitate a fox.

My backpack was loaded with canned beef dogfood, and although I felt somewhat strange littering the wilderness with this inexpensive meat, I was determined to parcel out this food in small artificial caches. Then in a week I would return and evaluate how the scavenging animals of this region had pilfered my stashes.

Experiment I: What Is All the Fuss About?

In all my field experiments, I mimicked a caching fox. At each station I dug a hole in the forest floor that was three or four inches deep, and then I dropped in approximately an ounce of dogfood and covered it with dirt. Instead of using my nose, I packed the dirt down using the blunt end of my fingers. Then I disguised the cache with leaves as well as I possibly could until I could not detect any trace of my buried cache. To be more precise, this is how I treated only half of my caches in this first experiment. For the other half, I dug a hole and left the cache

101

completely exposed—no dirt, no leaves, no twigs. I left the food in plain view, open to the elements. I wanted to see just how much difference all this fussing made in the preservation of the caches. Was all the trouble that the fox went to worth the effort?

To make the experiment as fair as possible, each time I made a cache I dug the hole and then randomly assigned whether it would be exposed or a camouflaged cache. I set out the caches in square plots, with 200 yards separating each cache and 4 exposed and 4 camouflaged caches to each plot. If I did not get my plots perfectly square in dense bush, such as the dog hair spruce I encountered, I reminded myself that the marauding animals would not care anyway. Once a plot was set up, I hiked a mile and a half further into the woods and set up another plot. I kept up my hoarding behavior until I had 60 camouflaged and 60 exposed caches scattered about the wilderness.

I resisted marking my caches with orange flagging tape because I felt the robbing animals might use this artificial material to find the food. Instead I made copious fieldnotes about the location of each cache and created small maps that led me from one hole to the next. I filled page after page with these jottings, and by the time I was finished I had all the more respect for the way a fox keeps track of its caches. I wondered if the neurons in a fox's brain are crammed with detailed notes or if the fox is blessed with a photographic memory of the forest floor. Whatever the mechanism, the fox's memory is impressive.

By the time I had all of my caches set out, the work had numbed not only my fingertips but also my brain. Only certain glimpses of the forest, such as a pileated woodpecker drumming solemnly on an ailing spruce, sustained me in this work. I had no way of knowing if I would be able to find these caches or if all of this work was for naught. But one week later I found myself standing where I had begun. I got out my notebook and began to follow my maps from cache to cache. My first delight was that I was able to follow my jottings from location to location, and with some additional searching I was able to look down into the same small hole I had made one week earlier and view the results of my experiment.

Working as a field biologist has its frustrations. Wild animals are free to wander off or die in the midst of your study. Weeks of rain can set in that make working and living in the field one of the most trying experiences on earth. But the work does on occasion have its rewards, and none is greater than to watch a pattern unfold as you collect the results of one of your experiments. This was my experience when I assessed the thievery to my caches: 60 percent of my exposed caches had been stolen but

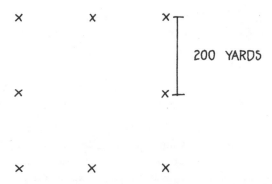

On each forest plot, artificial caches were set out 200 yards apart in a square matrix.

only 13 percent of my camouflaged caches had been pilfered. Clearly, it is not in vain that the fox is careful about its caches.

Furthermore, as I studied my vacant caching holes I often found clues that indicated who the thief had been. For example, at one cache a broad, shallow hole had been dug out, and a fresh bear track was authoritatively imprinted on the sandy soil. When a fox had dug out the cache, the excavated hole was narrow, about the width of two of my fingers; when a wolf had done the deed, the hole was wider, about as wide as my fist. At other caches the robbers inadvertently had left hair or feathers caught in shrubs found nearby the caching hole. All this evidence pieced together often suggested who the thief had been.

Every time I relocated a camouflaged cache that had not been disturbed, I wondered if the meat was really still present and what condition it might be in. I was unable to inspect these caches with my nose as I had often watched foxes do when they revisit and check on the welfare of their buried provisions. As a result, each time I found one of these caches I excavated the meat. I was repeatedly impressed with how fresh it appeared— almost as if it had been refrigerated the whole time. On the other hand, the exposed caches that had not been stolen were frequently being consumed in place by ants, fungal molds, and, most frequently, carrion beetles—those mysterious orange and black insects that appear out of nowhere whenever there is the smell of rotting meat.

From the first experiment, I concluded that the fox's labors are certainly worth the effort. The fox buries, packs, and camouflages its caches not only to protect them from pilfering mammals and birds but also from smaller thieves, such as scavenging insects and saprophytic plants, that devour the meat caches in situ. The experiment showed that it would be a foolish fox indeed who left its caches exposed in the wilderness.

Experiment II: The Fox as a Statistician or Why Doesn't the Fox Put All Its Eggs in One Basket?

I suppose if I had been wise I would have quit at this point, but a field biologist is rarely of such a prudent temperament. Excited by the results of my first experiment, I had already conceived of a more grandiose scheme and was eager to expedite it.

Animals that cache do so in one of two ways. The red fox *scatter hoards* its caches; that is, it places each load of food in a separate hole, and the holes are well spaced from one another. Most wild canid species that I am familiar with subscribe to this philosophy of caching. The other school consists of the *larder hoarders.* These animals, such as red squirrels and marmots, put all their surplus food in one protected hole usually close to their nest or burrow.

I have never seen a red fox larder hoard its food, nor has anyone else that I know of except J. O. Sande. During 1943 while living in Sweden Sande observed a red fox repeatedly putting prey into the same hole. When he examined the hole, it contained a hare, ten field mice, and a grouse. This is certainly more than one load of food for a fox. Normally, though, foxes scatter hoard their caches. The question is why. Why does a fox scatter its caches and subject itself to the considerable challenge of remembering where each is? What benefit is gained? I am not the first naturalist to wonder about this question; in fact it has been reflected upon for the last 200 years. The consensus among biologists seems to be that the fox cuts its losses by scattering its caches. Spreading caches out over the territory makes them harder for the robber animals to find, and thus the fox reduces the average number of caches lost to the competition.

I wanted to test this consensus to see if the scavenging animals of the boreal forest would support or reject it. Consequently I set up my second experiment in a completely different area from the first, a place where the animals had not been initiated to my tricks. Having psyched myself for the considerable effort I knew would be involved, I departed with a mother lode of canned dogfood in my backpack.

After hiking for a day and a half, I found myself standing on a small hilltop amidst a chapellike stand of white birch. I felt this place was an appropriately beautiful spot to begin so I reached into a pocket of my green fieldjacket, shuffled four nearly identical wood chips, pulled one out, and wrote "larder" on it. Then I proceeded to larder cache the meat. I placed nine chunks of dogfood, each weighing approximately one ounce, in the same hole and buried this meat under five inches of dirt. As in the first experiment, I packed this dirt down tightly with my fingertips (gloved this time) and proceeded to disguise the spot as skillfully as I could with leaves and litter.

I set up the next plot a mile and a half farther down the trail and several hundred yards off into the bush. I shuffled the remaining wood chips in my pocket, pulled one out, and wrote "3" on it. On this plot, each chunk of dogfood was placed in a different hole, and the holes were separated from each other by approximately three yards. As usual each cache was carefully buried, packed, and disguised, and this procedure was repeated until I had all nine caches set up in a tidy square plot.

Again, the same distance down the trail and off to the other side, I shuffled the remaining two wood chips in my pocket, pulled out one, and wrote "15" on it. I then set up a square plot of nine caches in exactly the same manner except that the caches this time were separated from each other by approximately 15 yards.

And again, down the trail and off to the other side, I pulled the last wood chip out of my pocket and wrote "75" on it. The same type of plot was set up, and each cache was treated in the same way except that the "intercache distance" was increased this time to 75 yards. My idea in all this was that if scattering the caches helped reduce the average number of caches lost to scavenging animals, then the more scattering the better. In other words, by progressively increasing the distance between the caches I should find a progressive decrease in the number of caches lost to the pilfering animals. At least this was the theory I hoped to test.

As with my first experiment, I did not use any flagging tape but continued to make copious notes and piratelike maps to lead me back to the caches after a week. At that time I hoped to see for myself if scattering saves caches.

One small point remained. I was talking about the *average* number of caches lost in each case. An average is unfortunately a statistical commodity. I had set up the experiment once, but to get a decent average I had to set it up at least 15 different times. It had taken me four hours to construct these original plots, and it looked as if I had a good deal of caching left ahead of me. However, there was no way around this brute reality; and so, blessed as I was with the almost continuous daylight of a northern summer, I spent the next several days working from 5:00 a.m. until midnight burying what seemed to be an endless supply of

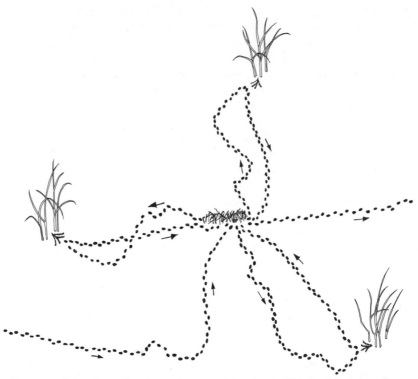

On a sand dune, tracks show that a red fox has robbed a herring gull's nest and cached the eggs. Characteristically the fox has scattered its caches, placing each egg in a different location (adapted from Kruuk, 1964).

dogfood. Each time I chose an area for caching, I shuffled the wood chips in my pocket and picked one out at random. This method assured that I randomly assigned the type of plot each area received. At the end of each run, I put four wood chips back in my pocket and began again. By the time I completed the last and four hundred and twentieth cache, I wondered why I had put myself through such an ordeal and how I was going to relocate these caches.

But after a week's rest, I was again standing in that beautiful birch forest, searching for my first larder cache. In the interim I had been thinking about all the meat I had stashed away under the forest floor, and I wondered if the robbing animals might not actually find the scattered caches, or at least a portion of them, more easily than the larder caches. After all, the widely spaced caches were spread out over a considerable amount of country, and it seemed likely that the robbers would stumble upon at least some of them. In fact, I had become totally confused about what to expect, and so I decided simply to collect the results and see what sense I could make out of the fox's predilection for scattering.

Once again the wild denizens of the area pilfered my caches with enthusiasm, and again the results began to roll in with the excitement of an election night. I got the feeling that the scavenging animals already knew the results of my referendum, and only a certain field biologist needed to be informed about the existence of this caching bylaw. The results tallied up this way:

Five of the fifteen larder-hoarded caches had been decimated— every bit of food had been cleaned out of these holes. The other ten larder caches were in perfect shape, well preserved and undetected by any scavenger. This amount of thievery averages out to two-thirds of the baits, or six out of nine, surviving the week. How did this compare with the other caching techniques? For the "3 yard" plots, an average seven out of the nine caches survived; for the "15 yard" plots, an average 6.47 out of the nine survived, and for he "75 yard" plots, an average 7.27 out of the nine survived. Lining up these averages—6.00, 7.00, 6.47, and 7.27— showed there was no consistent increase in survival as the caches were spaced farther apart. In fact, my statistical test told me that these averages simply fluctuated up and down and that no significant difference existed among them. Thus the experiment clearly rejected my hypothesis. But from personally having made 420 caches, I knew that it was considerably more effort to scatter rather than to larder hoard caches, and so once again I returned to my initial question: Why does the fox do it? What benefit does the fox gain from making this considerable effort?

I graphed the data first one way and then another trying to make sense of it. The experiment had rejected my hypothesis, and I was at a loss to come up with an alternative idea. The only idiosyncrasy I noticed was that when I larder hoarded my meat, I experienced an "all or none" type of loss. If a larder cache was detected at all by a scavenger, I lost everything. If the cache was not detected, my losses were zero. With this type of cache, my luck was either extremely good or extremely bad. How did this compare with my caches separated by 75 yards? Here my luck was more moderate. Usually a few of the caches were missing from each plot, but usually no more than a few. By scattering I had *not* reduced the average number of caches lost to the scavenging animals, but I might have made these losses more regular and uniform.

Perhaps I was onto something. No matter how the fox caches,

it seems destined to lose a certain percentage to the competition. Because the fox can do nothing about this tithe, the only question it can evolutionarily consider is: Can these losses be made more regular and thus more predictable? Or is caching necessarily a business of unpredictable food futures? To measure the variation among my losses, I calculated the standard deviations for each of the above averages. A standard deviation is simply a formula that measures how individual scores vary from the average. For example, I had already determined that my four caching techniques reported essentially the same average number of caches surviving after a week's time. However, for my larder caches the results of individual plots were highly variable (all or none of the baits were present); thus I would expect the standard deviation for this average to be fairly large. On the other hand, the losses for widely spread caches were more moderate; thus the standard deviation for this average should be smaller. After working with a calculator for an hour or so, my averages and their standard deviations looked like this:

Larder caches	6.00 ± 4.39
Caches scattered 3 yards apart	7.00 ± 2.83
Caches scattered 15 yards apart	6.27 ± 1.85
Caches scattered 75 yards apart	7.27 ± 1.67

What do these numbers mean? When I larder hoarded my caches, an average of six pieces of meat survived the week, give or take about four and a half. That is quite a lot of variation. When I scattered my caches 75 yards from each other, an average of about seven of these caches survived, give or take one and a half. In essence, the experiment showed that I had cut the variations in my losses to a third by widely scattering my caches—a real improvement.

The fox is a more sophisticated statistician than I thought it was. By scattering the fox regulates its losses. If the fox larder hoarded, there would be times when none of its caches would be stolen, but there also would be times when it was entirely cleaned out. Larder hoarding then would make the fox's caches an unreliable resource. By scattering the fox makes the loss of his caches more regular, thus increasing the chances that a portion of caches will reliably be there during a period of food shortage.

I had been well paid for my considerable efforts. A scientist enjoys no experience more than being told "no," or that something is more subtle than he expected. My initial hypothesis had been clearly rejected; scattering does not reduce the average number of caches lost to the competition. Instead it reduces the variance of these losses and thus makes the losses more predictable. The fox's caching behavior has indeed been sculptured to take advantage of these mathematical properties, and these subtle actions evolved millions of years before the words "statistics" and "variance" were invented.

Experiment III: Birds by Day, Bears by Night

Not all moments of inspiration come to a field biologist while he or she is in the wilderness. Two years passed and then one evening while I was among the refuse heaps at the national park dump a third experiment on the fox's caching behavior blossomed in my head.

I had realized for quite some time that the fox must defend its caches against two different types of animals: nose-oriented and eye-oriented scavengers. Bears, skunks, coyotes, mice, and foxes are examples of animals that may scavenge primarily with their noses while ravens, crows, magpies, jays, and gulls may rely primarily on their eyesight. In any case the fox must provide an olfactory camouflage as well as a visual camouflage for its cache. I began to wonder if this was the reason the fox's caching behavior was as stereotyped as it was. When the fox buries the cache and packs the dirt or snow over the top of it, perhaps the fox is creating a seal over the cache so that only a slight amount of food odor can escape. Once these actions are accomplished, then the fox begins to visually disguise the cache with leaves, twigs, and other forest floor litter. For humans, creatures with little smelling prowess, the visual camouflage is easily comprehended. Perhaps if we carried around the snout of a fox, the olfactory camouflage would be just as easily understood. These thoughts were interesting, but I was damned if I knew how to test them—until that notable evening at the Prince Albert National Park dump.

The way to understand an animal is often through its stomach. Where there is a preponderance of accessible food, there is likely to be an interesting collection of wildlife, and a dump is no exception. Within a national park, where animals have learned to lose much of their fear of man, a garbage dump with all its edible refuse can attract an interesting collection of animals—especially scavengers. Over the years I somewhat sheepishly have enjoyed my trips to the dump and on occasion have hung around for several hours to see who might appear. Wolves, wolverines, bears, and weasels are all among my prized dump sightings.

Make no mistake about it. My rational side would wish it all away if I could. Dumps do cause serious management problems within our national parks. Usually bears first become educated about humans and their abundant food supplies at dumps; later these same bears may begin to forage rather aggressively in camp-grounds. U.S. National Parks are making great headway on the proper handling of garbage within parks, and I hope that Canadian National Parks are, but I am referring here to a time before such enlightened management existed.

At any rate, the noteworthy evening at the dump I alluded to earlier was a particularly buggy July evening, and I was waiting expecting to observe black bears. After an uneventful hour, I was leaning back against a large rock looking through my binoculars when a young male bear ambled out of the woods, walked along a high ridge, and waddled down a sandy slope into the dump area. One second I was thinking how strange it was that bears always walked along that same ridge, and in the next instance my new experiment was conceived. That same ridge was the place where ravens, crows, magpies, and gulls congregated during the day when they were not picking over the garbage. Here was a location where I could observe birds by day and bears by night. On other evenings I had seen skunks, coyotes, and foxes also walk along this ridge on their way to the dump.

I thought to myself that if I set out caches along that ridge at dawn and left them there for eight hours then I could evaluate how the day shift of scavenging birds pilfered the caches. On the other hand, if at twilight after the birds had left I set out similar caches, left them in place for eight hours, and checked them at first light before the birds arrived, I could evaluate how the scav-enging mammals pilfered the caches at night. Would a visual camouflage be effective against the birds and an olfactory camou-flage be effective against the mammals? This is the hypothesis I wanted to test in my experiment.

My animal-watching at the dump now took on more serious overtones. I spent several days and large portions of the night observing the dump, verifying that my birds-by-day and mam-mals-by-night pattern was valid; indeed it appeared to be. Now I was ready. At sundown one evening, I crept out on the ridge, sang loudly to let the bears know I was there, and proceeded to make four different kinds of caches.

The first type of cache was left fully exposed. I dug a hole four inches deep, dropped in the dogfood, and left it uncovered.

The second type of cache was disguised but not buried. I dug the hole, dropped in the dogfood, and disguised the cache care-fully with leaves and litter.

The third type of cache was buried and not disguised. I dug the hole, dropped in the dog food, covered it, and packed the dirt down; but I placed no leaves or litter over this cache. From the freshly dug dirt and the impression of my fingertips in the soil,

Corvids, such as this magpie, as well as ravens, crows, and gray jays frequently follow red foxes that are carrying food. A fox responds to this unwanted surveillance in one of two ways: either the fox temporarily caches the food but soon after moves it to a new location or it simply continues to carry the food until the bird leaves. Corvids respond to the second tactic by quietly following the fox at greater and greater distances. In the case pictured above, a magpie attempts to rob the cache which the fox has just made.

there were visual indications that the soil had been disturbed.

The fourth and last kind of cache got the whole treatment—hole dug, dirt pushed in and packed down, and the spot disguised with leaves and litter as carefully as possible.

I continued my experiment during the next month. On selected days, I set up the experiment once at dawn and once at dusk. Each time I ran it I created 16 caches—4 of each kind—scattered along the length of the ridge. New locations were chosen for the caches each time, and the type that was assigned to each spot was randomly chosen—wood chips again. I continued this experiment for ten days and ten nights until I had set out forty of each kind of cache.

In one regard I found this experiment the hardest of all to execute. It was not the bears roaming in the bushes that disturbed me but the fact that the results were collected in a piecemeal fashion.

A pattern began to emerge in the results, however, and quite naturally I became excited about it. I then proceeded to expend considerable energy worrying and challenging myself, making sure that I did not prejudice the remaining trials of the experiment in any way. A scientist is always his own greatest skeptic; this is the way it should be, but at times this internal inquisition can be psychologically draining.

When the ten trials of the experiment were completed, I calculated the average number of caches lost to the scavengers. The animals elected to pilfer my caches with the following results:

	Average Number of Caches, Out of Four, that Were Lost to the Scavengers	
	Birds	Mammals
Exposed Caches	3.3	3.0
Disguised Caches	0.7	2.3
Buried Caches	1.4	0.6
Caches Buried and Disguised	0.1	0.5

It was now my task to figure out what this pattern of thievery meant. The first conclusion was easy: the exposed caches on the ridge did not last long. Within eight hours, three and often all four of them had been eaten. The results were the same during the day and at night. With fresh pieces of meat fully exposed, these results were not exactly surprising.

When I looked at the next type of cache, the results became more complex and interesting. In this experiment I assumed that when I disguised a cache without burying it I had removed most of the visual clues but left the olfactory clues readily detectable. What results did I get with this type of cache? On the average the mammals at night pilfered approximately two and a half of these caches, but the birds during the day got less than one. In fact, my statistics showed that considering the variances involved just disguising the cache was as effective in protecting it against the birds as both burying and camouflaging it. On the other hand, mammals found these caches almost as effectively as if they had been totally exposed. These results supported the hypothesis that the birds were locating the caches mainly from visual clues, and the mammals were locating them mainly from olfactory clues.

What did the next category show? Statistical analysis indicated that simply burying the cache without disguising it protected the cache against mammals as effectively as if it had been buried and camouflaged. Specifically the results showed that an average 0.6 of the buried caches were lost to the mammals as compared to 0.5 of the ones that were both buried and disguised—not much of a difference. Burial appeared to be the crucial element in a cache's olfactory camouflage, and camouflaging with leaves and litter appeared to contribute little to this olfactory camouflage.

On the other hand, the results showed that just burying the cache offered some visual camouflage against the birds (it reduced average losses from 3.3 to 1.4). However, burying was not as effective against the birds as was simply disguising the cache, which reduced average losses to 0.7 caches. But clearly the most effective protection of all against the birds was both burying and disguising; the pilfering birds found only one out of forty of these caches. Why does burying and disguising create the most effective protection against avian scavengers? A windy day on the ridge suggested a possible reason. When a cache had been recently disguised with leaves and litter, wind and rain might move around the leaves and twigs and thus visually expose the cache to the keen-eyed birds. Consequently I concluded that for a visual camouflage to remain effective through wind and rain the cache must be buried as well as camouflaged.

This experiment has its limitations; it was not executed in exactly a pristine environment. Nevertheless, it helps to explain why the fox keeps the burying and camouflaging actions of its caching behavior as separate and distinct as it does. From this experiment I concluded that burying and packing the dirt effectively hides the fox's caches from the talented noses that roam the forest. Once it is buried, camouflaging the cache with leaves and litter helps to hide the cache from critical eyes. Together it is an elegant system; one that helps to reduce unwanted teeth and beaks from dipping into the fox's pantry.

Take and Thou Shall Give

The most recent advancement in my understanding of the fox's caching behavior happened far from these animals; it occurred in a restaurant in downtown Edmonton. I was on a consulting assignment and had gotten together with another ecologist to reminisce about fine days spent in the field. I had recently completed my analysis of the fox's caching behavior, and so not unexpectedly I began to tell my friend about my field experiments—how I had set them up, what problems were encountered, and what the results had shown. I described the fox's main predicament; namely how the fox must bury its caches close to the surface so that if it cannot remember exactly where it put them it has a chance to locate them from olfactory cues. However, this shallow burial also increases the risks of having competing scavengers steal the caches. I listed the mammals and birds who rob the fox's caches and pointed out that one of the fox's worst thieves is of course other foxes. At this point my companion, who had been listening intently to all that I had told him, broke into my narration and asked, "Yes, but when that happens, when one fox steals another fox's caches, who is the fox doing the stealing?"

Who is the fox doing the stealing? It was a good question, an obvious one that I had somehow never thought of. It reminded me of a quote that I had read years ago and often reflect upon— that the inspiration of a scientist can be measured by the new questions he is able to ask. All this flowed quickly through my head. My friend had asked an insightful question about the fox's caching behavior, and I admitted to him that I did not know the answer. He said, "I have the feeling if you search out the relationship between the thieving fox and the fox whose cache is stolen, you may find something interesting." I thanked him and assured him that I would think about it and let him know what I came up with. Shortly thereafter he had to go to another commitment, but I sat in the restaurant thinking about his question. I remained there for a long while that evening. I was in the centre of Edmonton, surrounded by high-rises, but my mind was hundreds of miles away back in the boreal forest chasing after foxes.

I thought about that query not only for the rest of the evening but frequently over the next several weeks. My friend had asked a poignant question, one that had caused an avalanche of new ideas in my head about the fox's caching behavior. Each time I returned to it, some new idea seemed to fracture off and come tumbling down in my consciousness. It took a long while for me to pick my way through these boulders and to find order among this rubble of new ideas.

Red foxes occasionally show "food envy." This dog fox consumes a cache but threatens one of his offspring as it approaches.

I finally decided that when foxes rob one another's caches the most important question to consider is: Is this robbery harmful or beneficial to the foxes involved? On the surface it looks like simple competition among the foxes. They are competing for limited food, and so they steal one another's caches whenever possible. My friend's question, however, had caused a tremor in my understanding of the fox's caching behavior, and it had made me question if there might be a way that this robbery among foxes could be beneficial to all of the foxes involved.

The answer, I thought, seems to hinge around who is stealing from whom. What is the relationship between the fox who makes the cache and the fox who steals it? I recalled several radio-tracking studies that separately had found red foxes occupying well-defined family territories. Each territory was occupied by an adult male, one or several adult females, and their pups. Because foxes often organize their populations in this manner and because the territories are usually nonoverlapping, when one fox steals another fox's cache the thief may often be another member of the family. When this occurs, a stolen cache may not be exactly wasted. It may often be an involuntary contribution to the survival of the fox's offspring or mate. Consequently foxes may be a family of pilferers. I had seen nothing in the fox's behavior that suggested that a fox intentionally shared its caches. Everything that I had observed suggested that a fox is rather protective of its caches. These parsimonious tendencies had somehow caused me never to reflect upon who was stealing from whom. But once triggered I could see there was a way that this robbery could be put towards a good end—by keeping the thievery in the family.

My thoughts did not end at this point. After my next mental avalanche, I was able to find a second hypothesis concerning how the cache robbery could be beneficial to all. This theory suggests that the foxes, in taking any buried food regardless of whose it is, actually may end up retrieving a higher percentage of the caches tucked away on the territory. I had collected several observations over the years that showed that at least a portion of the fox's caches are wasted; that is, they are never used by any fox. Furthermore, the longer caches rest in the soil, the more likely they are to be stolen by scavengers other than a fox, and the less likely it is that the "owner" fox will be able to remember exactly where it put them.

I reviewed the different mechanisms by which a fox can retrieve a cache—remembering exactly where it is or recalling only the general area and locating the cache from olfactory clues. In addition, I had observed that a fox may simply stumble upon a cache—one that the fox itself had made and forgotten or, more likely, one made by a different fox. And finally, if a fox is on the trail of another fox whom he suspects is about to cache, the first fox may track that animal and locate the food from olfactory signals. The last two mechanisms often result in one fox stealing another fox's caches. If foxes were "honest," that is, if foxes retrieved only their own food caches, then the mechanisms by which they retrieve caches would be cut in half. Perhaps more caches would be wasted if foxes were "honest." By using all four retrieval mechanisms and stealing whenever they can, the number of caches wasted or lost to other species may be greatly reduced, and each fox ends up with more caches. Thus all the foxes benefit from this mutual robbery.

I admit that these ideas are speculative and that some of the assumptions contained in them can be challenged. I am still searching for an occasion to test these ideas in the field so that the foxes can either support or reject them. However speculative these thoughts are, they do remind me of how incredibly complex

116

the social life of a solitary animal can be. Rather than being simple or primitive, the social interactions of solitary animals can be just as complicated as the interactions of highly social species. If my days of fox watching have taught me one thing, they have taught me to see through the anthropocentric bias I once had when I thought that highly social species represented the pinnacle of evolution.

My friend's simple question had triggered quite a flow of thoughts. As I walked out of the restaurant through the concrete labyrinth of the city that evening, I wondered just how many more years it might take me to understand the fox's caching behavior—an action that takes the fox only 90 seconds to perform.

7 *Keeping Things Straight*

Mammals use urine marks for many functions: wolves and mountain lions, among others, mark the boundaries of their territories so that others of the same species (conspecifics) can tell that this area is already occupied. Many mammals (canids, felids, and certain rodents) use marking as a dominance display while other species, such as the polecat and slow loris, use the marks to follow trails. The common house mouse uses urine marks both to inhibit aggression between conspecifics and as an alarm substance. During sexual behavior male dogs, cats, bears, and males from many ungulate species use urine marks to locate females and evaluate their estrus status. Urine marks also help to provide that "long-distance feeling," which, within many canid societies, is a deposit of urine left behind to inform conspecifics about the sex, age, dominance status, and sometimes even the individual identity of the marker. Field research makes several points clear: first, mammals use urine marks for a diversity of functions; second, these uses vary from species to species; and third, each use needs to be confirmed through direct field experimentation.

After watching foxes for a few months, however, I began to doubt whether even all these accepted functions lumped together could explain the urine marking I was observing in red foxes. For instance, I was watching foxes deposit urine up to 70 times per hour while they scavenged for food. During these sessions the animal was not in the company of another fox, the breeding season was still five months away, and the fox was not scenting the boundaries of its territory but rather the entire area. I became suspicious that the foxes were urine marking for reasons that biologists had not as yet unearthed.

One point was clear: urine marking was one of the foxes' most common behaviors. Another characteristic I soon noticed was that foxes seldom released a large quantity of urine; instead whenever an animal squatted or lifted a leg it almost always released less than an ounce of liquid, yet foxes urine marked so frequently that they appeared to be able to eliminate all of their nitrogenous wastes through their scent marks. Only on rare occasions have I observed a fox perform a "normal" elimination, that is, release a large quantity of urine all at once. These noteworthy urinations were all performed by young, dispersing foxes—juvenile animals who had come onto another fox's territory and appeared to be trying to find something to eat before moving on. These foxes seemed inhibited about urine marking on another animal's territory. The few vigorous attacks I have observed territory owners make on transient foxes probably explains why they were reluctant to draw olfactory attention to themselves.

Because I had ample opportunity, I decided to apply the principle of exploratory watching to the fox's scent-marking behaviors; that is, I tried to put aside my preconceived notions of why animals deposit their scent and simply categorize the situations in which foxes marked, including the stimuli that seemed to elicit their behavior. After several weeks of classifying these actions, I began to detect some interesting patterns. First, I noticed that foxes investigate an object or spot on the forest floor prior to marking it, and that this investigation is a consistent feature of their urine-marking behavior. Next I learned that I frequently could not tell why a fox marked a particular site; the stimuli that the animal was responding to were often invisible to my senses. I assumed that if I had the olfactory acumen of a red fox, the reason for many of its urine marks would become clear. On the other hand, in some instances I was reasonably confident about why the fox had chosen to dampen a particular spot. For example, foxes seemed to have regular scent posts, that is, favored boulders, tree roots, shrubs, and other objects that they repeatedly urine marked. Every two days or so when a fox was passing one of these landmarks it would renew its scent on the location. This type of "social scent post" serves as communication in many canid species, and in the red fox it may also serve a territorial-marking function. However, objects that are repeatedly urine marked accounted for only 12 percent of the foxes' urine-marking behavior.

An additional documented pattern was the foxes' frequent use of urine marking while scavenging. When a fox found a carcass of a small animal or located some other food tucked away on the forest floor, it would usually eat the food, smell the spot thoroughly, and sometimes urine mark it. At other places, only the suggestion of food—a few feathers, a bone, some shed hair, or a dried patch of hide—would be urine marked by the fox before it left to scavenge elsewhere.

Because I had never found an explanation for why foxes urine mark so frequently during scavenging behavior, I decided to inves-

A dispersing juvenile male has crossed into territory occupied by an adult dog fox. The dominant male upon seeing the intruder immediately attacks him.

tigate. No insight was immediately forthcoming, so I decided to watch the foxes scavenge and mark for another month or two. Sometime during that time, it occurred to me that the fox might be keeping track of where it had investigated for food by urine marking those locations. I wondered if what I had been observing was the fox's "bookkeeping system."

I decided to try to write out precisely what I meant and found my thoughts somewhat muddled. In fact, my explanation evolved slowly over time with each revision accompanied by generous doses of exploratory watching. Finally I wrote:

> During scavenging behavior a fox stops
> and investigates up to 220 spots per hour on
> the forest floor, and a fox scavenges for

roughly 35 percent of its active time, or approximately five hours per day. Thus while scavenging a red fox investigates an extremely large number of spots on the forest floor, many more than it could probably remember without some external signal. Furthermore, I have observed the same fox, or several different foxes, scavenging in the same area several times within a two-day period. Thus urine marking may function as a type of "bookkeeping system" during scavenging behavior. The hypothesis is that a fox urine marks places where food has already been eaten but where food odor or inedible food remnants remain. When the same or a different fox reinvestigates this spot, the scent of fox urine signals "no food," and the fox will investigate this spot for only a short period of time. This use of urine marking might increase the efficiency of the red fox's scavenging behavior so that more food items are found per hour of scavenging. Efficiency of scavenging behavior may be particularly important during periods of food shortage.

This was an intriguing new hypothesis, but in some respects it seemed improbable—which of course only increased its scientific interest. However, I did not feel that it would be hard to disprove this conjecture. It seemed to me that the first point to attack was the contention that inedible food remnants could consistently elicit urine marking in the red fox. This point became the basis for my first experiment: Could things that left an odor or visual sign of food on the ground consistently cause foxes to scent mark these inedible remains?

I decided to test seven substances—gasoline, oil, synthetic urea, trimethylamine, water, soft dogfood, and dry granular dogfood— for their abilities to elicit marking behavior in foxes. The first four substances presented the fox with strong odors, but ones that were not food related. I chose these substances because N. Heimburger, a German researcher, had documented that the canid species he had studied under captive conditions occasionally investigated, rubbed, rolled on, and sometimes urine marked strong odors such as these. He suggested that these animals may be "decorating" themselves by rubbing against strong odors in order to stimulate conspecifics to investigate. I included water in my experiment as a control, because it is a liquid with a low amount of odor. The two kinds of dogfood were included to provide food cues, and they were used as follows: both were placed on the ground in small one ounce piles. When the fox ate the dry granular dogfood, few crumbs and very little odor remained on the ground. On the other hand, when the fox ate the canned beef dogfood the sign of food and odor did remain on the ground after the fox had eaten all available food. To ensure that this happened, every time I set out the soft dogfood I smeared a corner of it against a hard object (rock, tree root, or hard clay soil) immediately next to where I placed the food. This squashed portion, which the fox could not consume, created the "inedible food remnant."

To carry out the experiment, once every hour while I was following foxes I chose one of these seven substances at random and placed it on the ground a short distance from the animal. The fox usually approached, investigated the substance, and ate any food that was present. I then recorded whether the fox urine marked the substance. I realized that by occasionally putting food down I was reinforcing the fox's investigation of substances that I had placed on the ground; but I felt that I was rewarding the fox for taking part in the experiment, not reinforcing its urine marking behavior. The fox got the food whether or not it urine marked the spot. It also got an equal amount of food whether the food was granular or soft dogfood.

I continued the experiment for the better part of a month until each of these substances had been presented to the foxes 30 different times. Three different foxes (two adult males and a young female) participated in my experiment, and while these animals showed some variation definite trends emerged. Taken together the results of the investigations looked like this:

	Urine Marked	Not Urine Marked
Soft dogfood	23	7
Dry dogfood	4	26
Trimethylamine	1	29
Gasoline	0	30
Oil	3	27
Synthetic Urea	0	30
Water	1	29

One does not need a sophisticated statistical test to interpret these results. Clearly only one substance—the soft dogfood—was consistently urine marked by the foxes. They seldom urine marked the granular dogfood, any of the strong nonfood odors, or

A male red fox investigates a spot on the ground and then urine marks it. (The white arrow indicates the location of the same spot in both pictures.)

the water. Furthermore, the foxes rubbed against these substances on only four occasions. The soft dogfood—the only substance that left the odor and sign of food on the ground—was the only substance that consistently elicited urine marking by the foxes. My curious hypothesis had survived its first test, and it was at this point that I decided to take the speculation seriously.

A Smorgasbord of Stimuli

I had already thought out the next experiment and had a fair appreciation of how time-consuming it would be to execute. Nevertheless, I felt it would provide a reasonably good test of whether this "bookkeeping use of urine marking" was a real aspect of the fox's life.

For this experiment each day I selected a different area of undisturbed woods approximately 1,000 square yards in size. On the forest floor, I created five different types of spots in such a manner that each type presented the fox with an increasingly complex set of stimuli. Basically I wanted to test whether foxes would investigate for only a brief period of time sites where it detected the odors of both food and urine as compared to sites where food odor was combined with other stmuli (for example, a dug-out caching hole or the odor of a fox).

To conduct this test, I once again created artificial fox caches. I used this approach because I had often observed a scavenging fox locate a cache, dig it out, and eat the food; then the animal usually smelled the empty hole and sometimes urine marked it. More importantly I had frequently observed the same or a different fox scavenging within this area during the next several days. When the animal investigated the empty, urine marked hole, it abandoned it quickly and walked on. These observations provided a foundation for my second experiment.

The crux of the experiment concerned how the fox responded to these five types of spots. Type I spots were the artificial fox caches constructed in the usual manner. Type II spots were made by digging a cache-size hole and smearing a small amount of dogfood against a hard surface in the hole; these sites mimicked an empty caching hole where the odor of food was known to be present. If a fox came onto the experimental plot, investigated the spots I had created, and left without urine marking the site then it was reclassified a Type III spot. If, on the other hand, the fox urine marked the hole it became a Type IV spot.

Type V spots were different; they were places the foxes investigated that I had not manipulated in any way. The foxes frequently investigated these sites, but because they never licked, chewed, or dug at these places I assumed they never found any food there. Thus Type V spots were interpreted as places where the foxes were attracted by some feature but where no food stimuli appeared to be present. The following table outlines how the different kinds of spots were made and what stimuli were created at each location.

Type of Spot	How The Spot Was Created	Stimuli Presented To The Fox
Type I	One ounce of dogfood buried in 3-inch deep hole; hole filled with dirt and disguised with leaf litter	Odor of food
Type II	0.1 ounce of dogfood smeared on bottom of 3-inch deep hole	Odor of food Open hole
Type III	A fox investigated one of the above two types of spots and did not urine mark spot	Odor of food Open hole Odor of fox
Type IV	A fox investigated one of the above three types of spots and urine marked the spot	Odor of food Open hole Odor of fox Odor of fox's urine mark
Type V	Spot not manipulated in any way	No food was ever found at these spots. Foxes usually did not urine mark spots.

To carry out the experiment, I chose a new area of woods each morning and created 30 Type I and 30 Type II spots. The different sites were randomly mixed and scattered on the expanse of forest floor. Then I waited—sometimes for a few minutes, sometimes for hours—until eventually a fox came onto the plot and began scavenging. The fox usually dug out and ate the food contained in the Type I spots. It also investigated and often dug at the Type II spots even though it obtained little food there. With a stop watch, I timed how long the fox investigated each site and recorded whether the fox urine marked it. After the fox left, I created new sites until I once again had 30 Type I and Type II spots on my experimental plot. Usually after an hour or longer the same or a different fox came back onto the plot and began investigating again. I continued the experiment until twilight each day, and the next morning I set up the plot on a fresh area. The study was continued until the adult male and female fox that participated in the experiment had investigated each type of spot at least 30 different times.

As the results began to accrue, the first pattern I noticed was that these foxes responded essentially the same as the other foxes had done in the first experiment; that is, the male and female fox consistently urine marked empty caching holes containing the odor of food. In fact, they urine marked these inedible remains between 66 and 75 percent of the time. But their behaviors changed abruptly at the Type IV spots; at these empty caching holes previously marked by a fox they left their scent only an average of 17 percent of the time. There was a further significant decrease when we look at the Type V spots. At these locations the foxes urine marked only 1 percent of the time.

The next point that I looked at was how long the foxes investigated each type of site. The foxes' average investigation time plus or minus its standard deviation (based on a sample size of 30 for each timing) is presented in the following table.

| | Female Fox | | Male Fox |
	Trial #1	Trial #2	Trial #3
Type I spots	9.16 ± 2.97 seconds	7.35 ± 1.73	7.62 ± 2.28 seconds
Type II spots	6.82 ± 3.09	7.35 ± 3.36	6.68 ± 2.29
Type III spots	5.91 ± 4.15	1.45 ± 0.61	6.40 ± 4.45
Type IV spots	1.24 ± 0.51	1.22 ± 0.73	1.57 ± 0.57
Type V spots	1.75 ± 1.30		1.40 ± 0.62

What patterns do we see in all these numbers? The first point to note is that the foxes investigated the artificial caches (Type I spots) on the average of about 8 seconds give or take 2 seconds before they began eating. How does this compare with the other sites? The open caching holes containing the odor of food (Type II spots) were investigated essentially the same, an average of 7.0 ± 2.9 seconds, before the fox usually urine marked the hole and left to scavenge elsewhere.

Add the smell of fox to that open caching hole and not much changes; that is, the average investigation time for the Type III spots is 6.2 ± 4.3 seconds, not a significant reduction. But again things change sharply with the Type IV spots; add the odor of a urine mark to the hole and the foxes respond by investigating the spot for an average of 1.4 ± 0.6 seconds. It is apparent that even though there was the odor of food at these Type IV spots the added odor of a urine mark consistently elicited a short investigation time from the foxes. It is also clear that the open caching hole and the odor of a fox did not elicit this response; only the urine mark significantly shortened the foxes' investigation time.

The Type V spots help to suggest what the message of these urine marks might be. As mentioned earlier Type V spots were places that the foxes investigated but where no food stimuli appeared to be present. Foxes investigated these spots for an average of 1.5 ± 0.9 seconds, which is essentially the same amount of time they investigated the Type IV spots. Consequently the message content of these urine marks seemed to be equivalent to "no edible food present."

Let me stop now and consider whether this short investigation time could be due to memory. Does the fox remember where it had investigated and upon returning examine these places for only a brief period of time? This hypothesis does not explain all the results of the experiment. The female fox came onto the experimental plot more frequently than did the male, and she was usually the first fox on the plot each day. When the male came onto the plot and examined Type IV spots, I knew he was investigating places that had been urine marked by the female. He could not be using memory in these cases, yet he explored these Type IV spots only an average of 1.6 ± 0.6 seconds. Thus he showed the characteristic short investigation of these places even though he had not previously investigated them.

In summary, the results of this experiment support the idea that the fox does have a system to make its scavenging behavior more efficient. Specifically, whenever a fox encounters the visual sign or odor of food in a form that cannot be consumed the animal urine marks these stimuli. Upon reinvestigation the urine mark signals "no edible food present," and the fox investigates this remnant for only a very brief period of time.

Does the Fox Ever Ignore the Urine Marks?

The red fox appears to have an ingenious scavenging system, but I began to wonder if the foxes were ever misled by the urine marks. As a result, I decided to test how easy it might be to deceive a fox. I wanted to explore this question: If the amount of food odor at the site is artificially increased, would the foxes ignore the urine mark and carefully investigate the spot or would the message of the mark prevail?

To conduct this experiment, I made use of the services of The Prince, who I had observed but who had not participated in any of my previous experiments. I was able to follow this fox for long periods as he hunted and scavenged throughout the woods of his recently established territory. As I followed him, each time the fox urine marked the ground I carefully placed painted stones on either side of the mark so that I could precisely locate it at a later time. After the fox and I had become separated, I returned and randomly chose to bury one of two types of bait next to the urine mark. At half of the marks, I buried one ounce of soft dogfood in one of my standard artificial fox caches (three-inch deep hole, dirt packed over the bait, spot disguised with leaves and litter). At the other marks, I followed the same procedure except that only one-tenth of an ounce of dogfood was buried. In all cases I was extremely careful not to disturb the urine mark. Once everything was in place, I removed the painted stones and remembered the location from natural features.

In this experiment I was assuming that the larger bait would give off more food odor than the smaller bait, and I wanted to see what difference, if any, this would cause in the fox's investigation. I also assumed that the urine marks at these sites would be "active" for at least 48 hours. Thus if The Prince explored one of these manipulated spots within the next two days, I recorded his behavior. I soon found it was necessary to attract the fox to the vicinity of these sites, so whenever the animal was close to one I scattered a small amount of granular dogfood over the ground two or three yards from my buried cache. The fox usually walked over, ate the granules, and then often investigated the place where the bait had been buried. If the fox left without digging, I dug out the bait to verify that it was still there. Here are the results of this third experiment:

	Baits Dug Out	Baits Not Dug Out
1 ounce bait	24	6
0.1 ounce bait	2	28

Regarding the larger baits, The Prince on the average smelled the site 5.7 ± 4.0 seconds and then began to dig. He dug up 24 out of 30 baits and either ate the meat or carried it off to cache it elsewhere. Thus the fox was not "deceived" by the urine mark that was present; the animal responded to the strong odor of food instead of the odor of the urine mark.

Regarding the smaller baits, the fox smelled the area on the average of 2.3 ± 3.5 seconds, occasionally urine marked the site again, and walked off. The fox dug out only 2 of the 30 smaller baits.

This experiment shows that a fox's reaction to one of these urine marks is greatly dependent upon the amount of food odor that is present. In fact, there appears to be a hierarchy of stimuli that determines the fox's reaction to a spot. Its apportionment looks like this: strong odor of food > odor of a urine mark > weak odor of food. The highest stimulus present greatly influences how a fox reacts. For example, if the weak odor of food is the only stimulus present, the fox usually investigates the spot thoroughly. If there is both the weak odor of food and the odor of a urine mark, the fox usually investigates the spot only for a short period of time before scavenging elsewhere. Finally, if the fox encounters a spot where there is a strong odor of food plus the odor of a urine mark, the strong odor of food takes precedence, and the fox thoroughly investigates the spot. The fox seems to have its stimuli ranked and responds in an appropriate manner so that it does not miss any available food.

*   *   *

It is often said that a scientist should never "marry" or feel overly protective toward any of his hypotheses. If evidence goes against them, he or she should feel free to reject them and search elsewhere for a better explanation of the phenomenon. Furthermore, nothing in science is cast in stone. What seems reasonably well established one day can be disproven tomorrow by a more insightful and carefully designed experiment. Nevertheless, I conclude that, based on the available evidence, foxes do seem to have a type of inventory system; that is, they urine mark inedible food remnants or empty caching holes where the odor of food still lingers so that upon re-investigation, they waste only a brief amount of time on these unproductive sites. The only addendum I might add to this hypothesis is that while observing these animals I gained the distinct impression that if the food remnant was a particularly durable one, for example, a skull or long bone, the foxes tended to mark it with a more durable scent mark; they used feces instead of urine.

At present then the evidence appears to support the bookkeeping hypothesis, but it remains to be seen what light might be shed on it as a result of the work of other researchers. One of the most distressing yet productive dimensions of science involves the way in which other scientists challenge, prod, and sometimes wreak havoc with one's hypothesis or the design of one's experiments. It is therefore understandable that scientists sit somewhat on tenterhooks after putting forth a new supposition. Concerning the bookkeeping hypothesis, it is certainly too early to predict what will ultimately become of it, but it is encouraging that initial responses from several scientists have been moderately supportive. Shortly after I published a detailed description of these experiments (Henry, 1977), two researchers wrote to report that

they had used inedible food remains to elicit urine-marking behavior in their red foxes, and they essentially agreed with my
results. The hypothesis also received some support from Fred
Harrington, a biologist from Nova Scotia, who has studied the
relationship between urine marking and caching in two groups of
captive wolves. Harrington maintained these wolves in large,
fenced enclosures. Concerning their scent-marking behaviors, he
reports:

> Urine-marking never occurs when a cache
> is made, rarely occurs during later investiga-
> tions while the cache still contains food, but
> usually occurs soon after the cache is emp-
> tied. These findings parallel those for both
> red foxes (Henry, 1977) and coyotes (Harring-
> ton, 1979), suggesting that . . . the hypothe-
> sis can be extended to all three species. Ur-
> ine-marks deposited at caches signal the
> absence of food, permitting individuals to
> minimize time spent investigating previ-
> ously exploited sites. Further support for the
> hypothesis comes from the decreased time
> spent investigating empty caches after they
> are urine-marked, as shown by the wolves in
> this study. (Harrington, 1981: 284–85)

Harrington also found the situation to be somewhat more complex in wolves and coyotes. Among these animals only dominant
males and females within the social group urine marked the
exploited caches. Harrington also found with his coyotes and
wolves that individuals infrequently remark their own urine
marks on the emptied caches but usually remark another individual's. He suggests that "once a cache is urine-marked, the site
may take on additional functions, becoming an important social
focus where information on individual identity, reproductive and
social status, sex, *etc.* can be obtained" (Harrington, 1981:286).
At present this is where the matter rests. What happens to the
bookkeeping hypothesis in the future remains to be seen. As a
final note, let me say that I am struck by the similarity between
testing hypotheses and raising children. Both are initiated as a
result of creative moments. Both children and hypotheses need a
good deal of nurturing in order to develop, and ultimately both
must be let go of because in a very real sense they have lives of
their own to live. I suppose one more similarity must be acknowledged—in both cases the progenitor follows the development of
his offspring with considerable interest.

8   *New Foxes, New Ice*

*Yet here was the thing in the midst of the bones,
the wide-eyed, innocent fox inviting me to play,
with the innate courtesy of its two forepaws
placed appealingly together, along with a mock
shake of the head. The universe was swinging in
some fantastic fashion around to present its face,
and the face was so small that the universe itself
was laughing.*

Loren Eiseley
The Unexpected Universe (1969)

Because I have been trained in the science of animal behavior, I
analyze the empirical actions of animals; what foxes do—their
overt behavior—has been the focus of my research. As an ethologist I have also been taught to ignore the question of animal
emotions or awareness. I deal with the objective behavior of
animals because I have learned that if I limit my questions to an
animal's physical actions I can experiment with these behaviors
and often find answers to my questions. For example, I can take a
frivolous detail of the fox's caching behavior, experiment with it,
and find out if this seeming idiosyncrasy serves a useful purpose.
Or, having watched foxes trace arcs through the air, I can explore
the sculptured physique of a fox in order to understand the adaptations that have evolved that allow it to make impressive leaps
after prey. These things are possible because I have agreed to let
the empirical actions and physical features of the fox determine
the bounds of my research.

Yet my fundamental objective in studying red foxes was to
enter their world, their *Umwelt* as classical ethologists refer to it,
as completely as possible. I wished to cross over and perceive
things through their eyes, to enter their society, just as an anthro-

pologist sets out to live and become completely immersed in a foreign culture. But after years of study, I am not sure that my entrance into the realm of the red fox has been particularly successful or complete.

Although the research methods of ethology have been my principal resources, these techniques have also focused on the last major obstacle blocking my entrance into the province of the red fox. The hurdle I have yet to overcome is the issue of fox awareness—forbidden fruit for an ethologist. Let me be clear about this point; traditionally an ethologist does not deny that a fox may have emotions and its own brand of vulpine consciousness. Neither will he say that such things exist. He maintains that we have no way of knowing whether these things exist because we have no way of knowing what levels of awareness occur within the brain of a fox. Because no technique can be used to study these things reliably, an ethologist contents himself with the physical actions of an animal. If a question can be answered by examining the overt behavior of a species, a scientist will often take up the research challenge and try to find the answer. By contrast, if the question demands analyzing the animal's emotions or sentience, then he is likely to demur or to conclude that the question is beyond his grasp. Because the question exceeds the limits of his science, it is unanswerable.

This traditional approach may be completely adequate for studying other animals, but I have never found it entirely satisfactory for studying red foxes. To understand why a particular fox acts as it does, one must take into account its genetic inheritance, cumulative experience, and unique physical and social milieu. Only by attempting this type of synthesis can one begin to understand why an individual fox acts in the manner that it does. When I try to take into account the genetic variation found within the species, I find red foxes to be as individualistic as human beings. I do not mean to anthropomorphize the fox, to ascribe to it a human type of consciousness or emotion. This tendency must be carefully guarded against because it is always tempting to assume that an animal's motivation in certain situations is similar to what ours would be. I certainly do not wish to assign human qualities to an animal as different from humans as the red fox. On the other hand, I am not sure that I can concur with the traditional ethological viewpoint that fox awareness always remains beyond our reach.

Just as human reactions and understanding are often reflected in the physical actions we perform, I believe that as a result of certain peculiar situations foxes on occasion clearly exhibit a consciousness of the world around them. As a result of my field studies, I have become curator of some strange experiences with foxes, some fleeting impressions and nagging, unanswered questions concerning what kind of awareness occurs within the fox's head. Like the outline of a small fetal foot or elbow that presses against the wall of a pregnant woman's abdomen for a few moments and then disappears, giving the father a glimmer of the new person growing within his wife, foxes have shown me rough angles or "elbows" of their behavior that seem to indicate much about the life occurring within their bodies and brains.

One of these protrusions appeared the day I watched foxes react to new ice. Freeze-up in the northern woods comes suddenly, usually during early November on a windless and cold night. The lakes, which have been turning over for weeks, finally become calm, and a layer of cold, buoyant water forms on their surface. The water molecules begin to link up, and a layer of smooth, black ice begins to appear across the surface. By morning the layer is complete, and as it grows thicker the expanding ice groans and drums and makes odd, extraterrestrial sounds. My Cree friends say these sounds are the lake creatures calling for their warm blanket of snow—an expression of their ancestors' with no small ecological insight. But for a week or two, if the weather holds and the snow does not come, the lakes and ponds of the northern forest are covered with a special polished surface of dark ice. During the Novembers of several years, I have watched foxes react to this sheer ice. The young animals—the ones who have never seen ice before—are particularly interesting to watch.

I can still recall The Prince, dignified as he was, coming down to the lake early one morning and bumping his nose against the clear ice as he attempted to get a drink out of his accustomed lake. He was startled but then smelled the ice and began to explore it by waving a paw over it and making increasingly stronger scratching actions against its hard expanse. Slowly and cautiously the fox began to venture out onto the new surface, testing his way, carefully finding out if it would support all ten or twelve pounds of his weight. Within minutes his confidence in this odd new terrain seemed to grow as he walked gingerly at first but then more and more confidently across it. The fox continued, nose close to the ice, sniffing this new ground—a surface that oddly had few smells about it. Soon the fox resumed his normal trot, exploring with his nose as he went until one of his feet slipped out from under him in the midst of a normally coordinated trot. The Prince stopped, turned around, and as if puzzled smelled and looked at the slippery spot. He repeated the sequence several more times—trotting across the ice, slipping, and then stopping and investigating. Then suddenly he burst into a galloping sprint. Legs slipping and sliding everywhere, the fox for a moment was going nowhere; it was clearly a peculiar sensation for a normally graceful animal. The fox then raced across the surface and came to a sliding, half-stumbling stop, followed by a playful twisting leap into the air. Racing off again in a wheel-spinning start, sliding, sometimes falling onto his side, The Prince had, just like a young child, discovered the magic of ice.

From that moment on, ice-sliding sessions punctuated the next several days of this young fox's life. Periodically The Prince detoured from his scavenging and hunting to go down to the lake or a pond for a gliding bout. Now he took to the ice quickly, galloped in a semicontrolled fashion, and then slid—sometimes slipping down on his rump or crouching down low on his forelegs. The fox would race and skid back and forth across the ice surface for several minutes, and then he would be off again onto his more serious pursuits of procuring food during the last snow-free days of autumn. These play sessions continued until the first snows came, which of course blanketed the lakes and ended for all ice sliders—foxes and humans alike—a short but festive season that in the boreal forest marks the beginning of winter.

Over the years I have watched several young foxes discover the glaze of new ice. Their reactions were so similar that the whole episode has become somewhat predictable—surprise gives way to cautious curiosity that is gradually transformed into animated play. It is a delightful, joyous moment to witness in a fox's life, yet each time I see it, as a scientist, I am troubled because it is difficult for me not to believe that I understand what the young fox is feeling. If foxes and humans can respond to new ice in such a similar fashion, it makes me wonder how many other feelings I might share with a fox. It is one of those strange elbows that from time to time protrude momentarily out of the pregnant ethogram of the fox—a sign of the possible life within, a trivial behavior that leaves one wondering.

A few other aspects of the fox's behavior leave me pondering just how much the fox comprehends about the world as it unfolds around him. The fox's hunting behavior is a good place to look for these oddly shaped pieces left over from the supposedly completed puzzle of the fox. Consider, for example, a hunting technique I saw several different foxes use on rare occasions over the years. I call it "the nap and capture strategy."

Take Rose, for example, an adult female fox who in the spring of 1973 was raising two pups. Late one afternoon this vixen was carefully walking along the edge of a meadow, searching for prey in the shrub-covered slopes below her. At one place she sat and put her head in mousing position. She peered at the area in front

134

of her for several minutes, periodically rotating her head around the axis of her neck, looking and listening first from one angle and then another. Then slowly crouching down in front, she lunged two yards down the slope and partially disappeared amidst tall grasses and shrubs. She bit down, raised her head, searched intently for a moment and then, just as quickly, marched out of the bushes and back to the top of the hill. Then she did something that I had never seen a hunting fox do. She circled around herself, curling up as foxes usually do when they are going to lie down, and she did just that, positioning herself so as to face the spot where she had just hunted. She put her head down on her soft, bushy tail, shut her eyes, and appeared to sleep. I stood stock still not wishing to disturb her. She remained like this for 12 minutes until some rustling sounds in the shrubs in front of her caused her to become alert. Slowly rising to her feet, she stalked closer to the shrubs. She then raised her head into mousing position and peered at a spot. Then without making a sound she crouched deeply and lunged downhill, this time pinning her prey

A fox naps but also watches the entrance of a burrow for a mouse to reemerge.

solidly to the ground. She bit several times at the animal pinned under her forepaws and then picked it up in her mouth and brought it up onto the grassy hilltop. For several minutes she played with the rodent (a *Microtus* vole), tossing it up in the air and catching it in her mouth again before laying the lifeless prey on the grass in front of her. She then picked it up, lifted her mouth above her head, and began to chew the small cadaver with her carnassials; I can still recall the muffled sound of small bones snapping as she chewed. Then after 10 seconds Rose swallowed the vole, skin, fur, and all, and set off searching for other prey in the surrounding woods.

The first time I saw this "nap and capture technique," I had a lingering doubt about what had occurred. Perhaps it was just coincidence. Perhaps the fox hunted and missed the prey, then by chance decided to take a nap. The prey became active again, and the fox was given another chance at capturing it. After I had seen several foxes use this slumbering hunting strategy, however, my doubts about what was actually taking place all but disappeared. Rose's hunt describes how it usually occurs. Rose seemed to be aware that the prey had vanished into a burrow so she lay down within lunging distance of the hole, carefully facing it. From her actions she seemed to be aware that there was a chance that the prey might appear above ground again. She also seemed to understand the importance of not making any sounds or movements so she napped. The fact that Rose did not continue her nap after she had caught the vole and did not nap during the next several hours further suggests that in this case her napping had other purposes than just resting. Although it is hard to make any absolute decision, there appears to be more purpose about this behavior than mere coincidence. It is another one of those protuberances in the fox's behavior that cause me to wonder just how much understanding of mice, voles, and other prey a fox carries around inside its head.

Other instances also lead me to think about this point. I flip my field notes over until I am back in the fall of 1971. After searching the pages, I find a hunting episode involving My Friend, the skillful male fox who lived on my study area for seven years. This incident happened when My Friend was a young fox, only six months old, and occurred after he had by chance been successful in three hunts in quick succession. The first two mice the young fox killed he immediately ate; the next mouse the fox played with vigorously for several minutes. When the mouse finally died, My Friend carried it off a distance and cached it. My field notes refer me to Hunt No. 56; I sort my pile of index cards until I find this one, and then I read about the hunt:

My Friend continues to hunt in this area
of woods. He walks down the path searching
for prey on the downhill side of the trail,
and all of a sudden he lunges again and,
remarkably, is again successful. This hunt
happens very quickly, just after he finished
caching up the prey from the previous one.
This present hunt is a successful shrew
hunt. He hears something in front of him
and runs a couple of steps at it, dropping his
shoulders and holding his head and neck low
and horizontal to the ground. Then without
pausing he made a short lunge and bit down
and captured the prey. Then he turned and
climbed the hillside carrying the prey in his
closed mouth up and out onto the roadway
where he set it down and began to play with
it. I get the impression that he knows that
the road is a safe surface upon which to play
with prey—there are no burrows on the road
surface to aid in a prey's escape. My Friend
is up there now playing with the shrew; the
fox is leaping around, dancing about the
shrew who runs over to one side of the road
before the fox herds it back to the center.
After 45 seconds of playing with this animal,
the fox then does an extraordinary thing. He
picks the shrew up in his mouth, walks back
down the slope to where he captured the
prey, and then with a toss of the head spits
the shrew out directly at a small burrow. In
less than a second, the shrew disappears into
the hole and is out of view. The fox watches
the shrew vanish and then casually turns,
trots up to the roadside again, and begins to
walk down the road searching for other prey
in the long grasses that grow along this roadway. . . .

The behavior shown by My Friend left little doubt in my mind
that the fox had intentionally released his prey. Over the years I
had observed other foxes casually stand and watch captured prey
scurry away from them although this did not happen often. In
these other occurrences, however, there was some room for sus-
pecting that these prey had managed to escape—but not in this

incident. The fox had carried the prey back to where he had captured it and flicked the prey directly at a burrow—an escape hatch only several inches away from where the shrew had originally been caught.

Over the years I have often thought about this episode. What seems clear to me is that the fox chose to let his prey go. He did not accidentally open his mouth so that the prey fell out; the fox tossed it at the burrow, and the shrew disappeared into its subterranean retreat. Shrews are admittedly not the preferred food of the red fox; the foxes I observed tended to cache these prey more often than they ate them. Some researchers have gone so far as to suggest that shrews are unpalatable to foxes, but on a number of occasions in my area I have watched foxes kill and consume these tiny prey.

After reflecting on this episode, one of the few things I am certain about is that the fox could have killed and cached the prey; he had done this just moments before. Yet this young fox chose to release the shrew. Why? To me it seems reasonable to say that under these circumstances the fox valued the shrew more as an item to be captured again than as a food item. If the prey had been valued as food, he would have killed and cached it. But the fox put the shrew back, perhaps to have the enjoyment of catching it again.

What is more difficult for me to reckon with is the sense—at least the primordial sense—of understanding about the shrew that My Friend seemed to exhibit as a result of his behaviors. The fox carefully picked up and delivered his prey back to where he had caught it. Does this in any way indicate that the fox understands that the shrew is alive, and that it has a home area just like a fox? By tossing the shrew directly at the burrow, does the fox show some sort of understanding of the importance of that burrow to the future existence of the shrew? If not, then what is a better, simpler explanation for the fox's actions?

I may never know the answers to my questions, but the episode leaves in my mind some interesting doubts about the kind of predator the red fox is. One traditional view of predation proposes that it occurs when instinctive motor patterns are triggered and released by the proper stimuli. For example, the cat hunts any small, quickly moving object whether it is mouse, bird, or small ball of yarn. The innate stalking and attack motor patterns are released by the proper key stimuli; it happens automatically—fox and cat kill their prey because these inborn hunting patterns have been triggered. They have little control over their actions and are indeed "innocent killers." If nothing else the episode with My Friend suggests that a lot more might occur inside a fox's head

Many features of the fox's strategy for hunting small burrowing mammals seem directed toward minimizing the amount of noise made during the hunt. Here a fox quietly listens for prey to move under the snow.

when it is hunting than this traditional view of predation allows.

It would be easy to ramble on philosophically about the meaning of this incident between the fox and the shrew, but I am not sure my conjectures would have much scientific justification. On the other hand, it would be easy to ignore this incident entirely— to pass it off as a fluke. Perhaps that is not scientifically justified either. If we are going to understand red foxes completely, in all their depth and breadth, we cannot study and analyze just their common behaviors—their everyday actions that are often repeated, easily described and measured, or just those behaviors that can be quantified. I believe that we can learn as much about animals from studying rare events in their lives as we can from analyzing their basic actions. The fox's rare behaviors, the ones that are difficult to observe and often hard to interpret, may give us valuable clues for comprehending the types of awareness and understanding that occur inside an animal's head.

If we choose to ignore these rare but important events—these "pregnant elbows" that stick out of a fox's psyche from time to time—then we can be assured of only one thing: that we will end up with a biased and possibly erroneous understanding of the red fox. If we study only common, convenient behaviors, then we will end up with a stilted and narrow understanding of what it means to be fox—an animal vying for its existence, a magically nonrandom event unfolding in one obscure corner of an expanding universe.

Notes

Page
Preface JET-AGE FOXES
22   Vegetation, climate and geomorphology of Prince Albert National Park and surrounding area. See Rowe (1959), Aldrich (1963), Swan and Dix (1966), Jeglum (1972, 1974), and Gimbarzevsky (1971).

1 IN THE COUNTRY OF THE FOX
25   Quote taken from N. Tinbergen. 1966. Adaptive features of the black-headed gull, *Laurs ridibundus* L. In: D. Lack (editor) Proceedings of the XIV International Ornithological Congress, p. 43–59. Blackwell Scientific Publications. London (reprinted by permission).
25   Predator/prey cycles in the boreal forest. See, for example, Keith (1963, 1974).

2 THE MASTERFUL FOX
27   Phylogeny of the red fox. See Romer (1966), Colbert (1980), and Radinsky (1973, 1981, 1983).
29   For detailed reviews of the red fox's food habits, see Korschgen (1959), Englund (1965), Goszczyński (1974), Jensen and Sequeira (1978), Pils and Martin (1978), von Schantz (1980), Hockman and Chapman (1983), and other references listed in the bibliography.
30   Macdonald (1977) documented that food preference among individual foxes can change in certain circumstances such as when rearing young or when competing for food (see also Jones and Theberge, 1983).
30   Lindström (1983) found a correlation between the availability of berries and fruit and the build up of winter fat reserves in Swedish red foxes. The degree of depletion of these fat deposits during winter was correlated with mean snow depth.
30   Red foxes, consumption of avian species. See food habit studies cited above plus Kruuk (1964), Norman (1971), and Sargeant (1972, 1978).
31   For a more detailed description of the world distribution of the red fox, see Fox (1975), Walker (1978), and Lloyd (1980a, b).

Page

31   Ironically hunting has led to an increase in the worldwide distribution of the red fox. For example, in 1868 for the purpose of fox hunting British red foxes were imported and released in Australia. Over time the species has become well established on that continent (Troughton, 1957). In North America, before Europeans arrived, the red fox may have had a more restricted distribution (see Churcher, 1959, 1960). However, British red foxes were also released in North America during the late 1600s, and logging and agriculture opened the eastern forests and created more productive red fox habitat.

31   Taxonomic and ecological comparisons between different species of foxes. See, for example, Thornton et al. (1971, 1975), Henshaw et al. (1972), Rohwer and Kilgore (1973), Fox (1975), Stains (1975), Clutton-Brock et al. (1976), Hersteinsson and Macdonald (1982), and Hockman and Chapman (1983).

32   Wolves as predators. See Mech (1966, 1970), Pimlott (1967), Ewer (1973), Peterson (1977), Allen (1979), Carbyn (1974, 1981), and Harrington et al. (1983).

33   Activity patterns of red foxes. See Tembrock (1958a), Ables (1969b, 1975), Allison (1971), Henry (1976), and Kavanau and Ramos (1972). An interesting speculative discussion of activity patterns is given by Meddis (1975).

33   Experimental determination of red fox auditory capabilities. See Österholm (1964), Peterson et al. (1969), and Isley and Gysel (1975).

34   Productive red fox habitat. For a more detailed analysis, see Ables (1975) and Lloyd (1980a,b).

35   Red foxes' preference for edge environments. See Ables (1975), Henry (1976), and Zimen (1980a).

36   For an analysis of the factors that determine carnivore home-range size, see Gittleman and Harvey (1982) and Bekoff et al. (1984).

36   Englund (1970, 1980) compared the relationships among food supply, productivity, population density, and dispersal patterns in red fox populations in a southern and northern area of Sweden (see also Lindström, 1980). He concluded that in the areas where food supplies were fairly stable population density of the foxes per se was an important factor in regulating their numbers; but in the area where food supplies fluctuated red fox numbers were limited directly by food availability.

36   Red fox families occupying nonoverlapping territories. See Tembrock (1957a), Sargeant (1972), Macdonald (1979b, 1980d), and Niewold (1980).

37   Small amount of sexual dimorphism in red foxes. See Fairley (1970) and Storm et al. (1976).

37   The social organization of the red fox can be contrasted with that of the coyote (Canis latrans) by referring to Bekoff and Wells (1980, 1982, 1985), Bowen (1981), and Messier and Barrett (1982).

 (Johnston and Beauregard, 1969). Laboratory and field studies have shown that red foxes are fairly susceptible to the disease, although there may be a long latency period between when the fox is first infected with the virus and the appearance of visible symptoms (Macdonald, 1980c and Zimen, 1980a). This latency period probably explains why some writers in the past have erroneously claimed that the red fox can carry and transmit the rabies virus but has a natural immunity to the disease.

In Europe the spread of rabies has been shown to be transmitted largely by red foxes; that is, through social encounters among adult foxes or the dispersal of young foxes into new areas (see Zimen, 1980a for a recent review). Field studies in Germany have shown that when fox population density is reduced below 0.2 fox/km² through hunting and the gassing of fox dens, the frequency of rabid foxes drops significantly and the geographic spread of the disease is controlled as well (Bögel et al., 1976). However, Zimen (1980c) argues that over the long term these control measures may actually catalyze higher reproduction rates among vixens as well as stimulate a higher percentage of young foxes to disperse from their parents' territories. A promising new control technique is the innoculation of wild foxes against the disease through the use of meat baits carrying an effective rabies vaccine (Bacon and Macdonald, 1980). This approach does not disrupt the natural population structure of the red fox; consequently in areas where food resources are stable fox reproduction and dispersal rates may remain low (see Macdonald, 1980c; Zimen, 1980a; Steck et al., 1982 for detailed discussions of this technique).

143

3 GROWING UP FOX STYLE

Page

44   This chapter was first published as an article entitled "The Little Foxes," *Natural History* 94 (January 1985): 46–57. It appears here in a slightly modified form and is reproduced by permission of the editors of *Natural History*, The American Museum of Natural History, New York.

44   Some interesting observations on fox courtship are provided by Taketazu (1979). In this pictorial volume, many aspects of the life history of red foxes are captured in handsome photographs.

44   Dates of breeding season in the red fox and how they change with latitude. See Ables (1969a, 1975) and Storm et al. (1976).

44   The chemical composition of red fox urine marks during the breeding season. See Jorgenson et al. (1978).

44   Resident male–female pairs of foxes defending territories. See Murie (1936, 1961), Barash (1974), Preston (1975), Henry (1976), Macdonald (1979b), and Niewold (1980).

45   Characteristics of den sites. See, among others, Sheldon (1950), Allison (1971), and Storm et al. (1976), Pils and Martin (1978), Stubbe (1980), and Macdonald (1980d).

46   Number and types of dens used by a family of red foxes. See Sheldon (1950), Layne and McKeon (1956a), Sargeant (1972), Storm et al. (1976), Pils and Martin (1978), Savage and Savage (1981).

46   Conception-birth dates (gestation periods) and litter size of red foxes. See, for example, Storm et al. (1976), Pils and Martin (1978), and Zimen (1980a).

47   Layne and McKeon (1956a) describe the size and development stage of neonatal fox cubs.

47   Charcoal grey pelage of newborn red fox cubs. See Taketazu (1979).

48   Infantile characteristics eliciting affectionate responses in humans. See Lorenz (1943) and Gould (1979).

48   Henry (1976) and Macdonald (1980d), among others, describe how the vixen closely attends the kits during their first seven to ten days of life.

48   Dog fox bringing food to vixen. See Sargeant and Eberhardt (1975), Henry (1976), Macdonald (1980d), and Niewold (1980).

49   The establishment of a strict hierarchy by pups during early life is described for the red fox (see Bekoff, 1974, 1978) and for the coyote (see Bekoff, 1978, 1981). For mortality patterns among young red foxes, see Dekker (1976).

51   Occasional predators of red foxes. See, for example, Allison (1971), Dekker (1976, 1983), and Voigt and Earle (1983).

51   Weaning process in red foxes. See Allison (1971).

52   Communication displays of the red fox. See Tembrock (1957a,b, 1958b, 1959), Fox (1969b, 1970, 1971b), Henry (1976), and Macdonald (1980d).

52   Role of dog fox in raising the kits. See Seton (1929), Ables (1975),

Page

Henry (1976, 1985), Macdonald (1980d), and Savage and Savage (1981).

53   "Helper" females may contribute to the successful rearing of a litter of red foxes by helping to defend the family territory against foreign foxes, bringing food to the pups, playing with and grooming the pups, and on rare occasions even nursing the pups (see, for example, Macdonald, 1979b).

53   The role of "helpers" is described in detail for blackbacked jackals (Moehlman, 1979), coyotes (Bekoff and Wells, 1982), and brown hyenas (Owens and Owens, 1984). This phenomenon of alloparental behavior as observed in other carnivore species is reviewed by Bekoff et al. (1984).

54   For a discussion of reproductive restraint in young red fox vixens, see Macdonald (1980d) and Macdonald and Moehlman (1982).

55   Conflict between parent and offspring. See Trivers (1974) and Wilson (1975).

55   Juvenile foxes waiting for long periods for adults to arrive with food. See Allison (1971) and Henry (1976).

57   Storm et al. (1976) offers detailed analysis of red fox dispersal patterns; see also Englund (1970, 1980b), Phillips et al. (1972), Fox (1975), Pils and Martin (1978), Lloyd (1980a), Schantz (1981), and Zimen (1984).

4 FOX HUNTING

58   This chapter was first published as an article entitled "Fox Hunting," *Natural History* 89 (November 1980): 60–69. It appears here in a somewhat expanded form and is reproduced by permission of the editors of *Natural History*, The American Museum of Natural History, New York.

59   The hunting strategies and concomitant morphological features of various canid and felid species can be confirmed by referring to the following references:

Murie (1936, 1940, 1944, 1961)
Scott (1943)
Leyhausen (1956, 1965a,b, 1973)
Hildebrand (1959, 1961)
Gauthier-Pilters (1962, 1967)
Egoscue (1962)
Kruuk (1964, 1972)
Kühme (1965a,b)
Mech (1966, 1970, 1975)
Schaller (1967, 1972)
Kruuk and Turner (1967)
Estes (1966, 1967, 1974)
Estes and Goddard (1967)
Pimlott (1967)
Pimlott et al. (1969)

Ewer (1968, 1969, 1973)
Walther (1969)
Eaton (1969, 1970, 1974)
Hornocker (1970)
Gray (1970)
Lawick-Goodall (1970)
Koenig (1970)
Speller (1972)
Kuyt (1972)
Carbyn (1974, 1981)
Lawick (1974)
Malcolm and Lawick (1975)
Chesemore (1975)
McMahon (1975)
Peterson (1977)
Bekoff and Wells (1980, 1982, 1985)

The points presented here are a summary of a detailed discussion found in Henry (1976); see also Kleiman and Eisenberg (1973).

60 The coyote is sometimes found to be solitary, other times to exist in small packs. Bekoff and Wells (1982) conclude that food supply is the major ecological variable influencing juvenile dispersal and coyote social organization. For example, when carrion is abundant, clumped, and defendable, packs of coyotes are observed; when carrion is scarce and spread out geographically, resident-mated pairs and solitary individuals predominate (see also McMahon, 1975; Bowen, 1981; and Messier and Barnette, 1982).

63 Vulnerability of ground nesting birds to fox predation. See Kruuk (1964), Englund (1965), Norman (1971), and Sargeant (1972, 1978).

64 A more detailed description of hunting strategies that have been observed in red foxes of the boreal forest can be found in Henry (1976).

64 Prey-capture techniques shown by the red fox are also described by Murie (1936), Tembrock (1957a,b), Kruuk (1964), Fox (1969a), and Savage and Savage (1981).

64 To compare the hunting strategies used by red foxes with those used by coyotes, see Bekoff and Wells (1980, 1985).

65 Red foxes as predators of earthworms. See Macdonald (1980a).

66 Head cocking by red foxes or other predators to locate prey—see Goethe (1950), Österholm (1964), Morrell (1972), and Konishi (1973).

67 Length of coyote's hunting lunges (Murie, 1940; Bekoff and Wells, 1980, 1985; Bekoff, pers. comm.; and Henry, unpublished data).

68 For a detailed description of the behaviors by which a red fox captures black-headed gulls *(Larus ridibundus)*, see Kruuk (1964).

70 Diagnostic characteristics of Canidae. See Simpson (1945), Stains (1967, 1975), Clutton-Brock et al. (1976), Radinsky (1973, 1981), and Scapino (1981).

also Powell, 1974 and Hamilton, 1971).

77   Experiment with white-footed mice and screech owl. See Metzgar (1967).

78   Maze experiment. See Kavanau (1967).

78   For an interesting study of death-feigning behavior in ducks captured by red foxes, see Sargeant and Eberhardt (1975).

79   Compensatory reproduction. See Errington (1967).

82   Formula for range of a projectile. See Spiegel (1967).

86   Trees tend to grow parallel to the force of gravity. See, for example, Horn (1975).

87   Analysis of the gait of saltatorial animals. See Gambaryan (1974) and Hildebrand (1974).

87   Gambaryan (1974) analyzes the muscular and skeletal adaptations that many mammals exhibit for running and concludes that the red fox is moderately specialized for fast running. He also gives the relative weights of the forelimb muscles of the red fox as compared to these weights for other felid and canid species. Unfortunately for my study, Gambaryan did not analyze whether the red fox shows any specialized adaptations for lunging.

87   Analysis of body proportions of mammals. See Hildebrand (1952a, 1954, 1974), Davis (1964), and Gambaryan (1974).

87   Quote taken from p. 234 of M. Hildebrand (1952), An analysis of the body proportions in the Canidae. *American Journal of Anatomy* 90: 217–256.

88   Size and weight comparisons between male red foxes from Ireland and female coyotes from the American Southwest. Mean ± standard deviation is given for each measurement (data from Fairley, 1970, and Young and Jackson, 1951).

	Red Fox	Coyote
Head and body length (inches)	28.5 ± 0.2	29.7 ± 1.7
Body weight (pounds)	15.2 ± 0.4	26.5 ± 3.1
Sample size	n = 42	n = 12

88   Suggested show weights for different breeds of dogs. See Fiorone (1971).

89   Stomach weights and feeding capacities of wolf and red fox. See Mech (1970) and Lloyd (1980a).

89   Relative width of canid skeletal elements. See Hildebrand (1954).

91   Osteopenic properties of bird skeletons. See Thompson (1942), Van Tyne and Berger (1959), and Romer (1962).

91   Taylor's technique is critically reviewed regarding allometry and other properties by Davis (1964); see also McMahon (1973) and Hildebrand (1974).

6   CUTTING YOUR LOSSES

95   "Cutting Your Losses" is a revised version of a scientific study entitled "On the Caching Behavior of the Red Fox: The Adaptive

Value of Camouflaging and Scattering Caches" (Henry, 1976). The reader should refer to the scientific study for a fuller description of the three experimental designs, for a statistical analysis of the results, and for a discussion of the limitations inherent in each experiment.

96 During my field studies (see Henry, 1976), I was able to document that the caching behavior of red foxes varies systematically according to at least two different factors: (1) "value" of the food item being cached (meat versus nonmeat food items) and (2) the onset of inclement weather. Thus foxes tended to cache meat food items more carefully than they did nonmeat items (fruit, vegetables, or bones cleaned of meat). Foxes also tended to cache more carefully during the onset of wind, rain, or snow-storms than they did immediately prior to or after these storms. "Foxes cache more carefully" means (1) they carried the food farther, (2) investigated significantly more potential sites before selecting the actual location for the cache, and (3) although the hole dug for the food item was not any deeper, the foxes spent significantly more time burying the food, packing the dirt, and disguising the cache with leaves and litter (see Henry, 1976). Macdonald (1976) reports related findings. In his leash-walking experiments with captive foxes, a particular vixen became rather slovenly about burying food and covering it with debris if she had been well fed for several days. But having one of her caches looted by crows or depriving the vixen of food for 24 hours restored her careful caching behavior. Macdonald also found that his foxes cached preferred prey (e.g., field voles, *Microtus agrestis*) more consistently than less-preferred prey (e.g., bank voles, *Clethrionomys glareolus*).

97 Macdonald (1976) suggests that a red fox under certain circum-stances may cache while still hungry, for example, as a way of protecting food from competitors or of increasing the time avail-able for prey capture.

97 For estimates of food requirements of red foxes, see Lockie (1959) and Sargeant (1978).

97 Exploratory watching. See Tinbergen (1951, 1962, 1963a,b, 1965, 1969a).

97 Evolution of behavior by natural selection. See, among others, Cullen (1957), Tinbergen (1963a,b), Crook (1965), Ewer (1968a), Brown (1975), and Wilson (1975).

98 See Macdonald (1976) for a review of studies that document that red foxes often retrieve their caches.

98 Red foxes' foraging activities around the Ravenglass gull colonies. See Kruuk (1964), Tinbergen (1965, 1966), and Ennion and Tinber-gen (1967).

98 For experimental evidence of the ability of red foxes to recall the precise location of caches, see Macdonald (1976). He concludes that memory is the most important mechanism by which red

Page

foxes retrieve caches. On the other hand, my field studies, while they agree that memory is an important mechanism, also suggest that red foxes have other mechanisms by which they locate caches. Specifically my field observations suggest that red foxes may sometimes only remember the general area where the cache is and detect its specific location from olfactory and visual cues, or a fox may locate a cache by tracking another animal who seems likely to make a cache, or the fox may detect a cache from olfactory or visual cues as part of the fox's general scavenging activities (see also Murie, 1944, 1961; Kruuk, 1964; and Tinbergen, 1965).

98 Scavenging animals that rob the caches of a red fox. See Murie (1936, 1961), Scott (1943), Tinbergen (1965), and Macdonald (1976).

98 During the course of my field work, I observed freeranging, wild red foxes "steal" a cache made by another fox on more than 40 different occasions. Thus in my area, "pilfering" another animal's caches was a reasonably common practice among red foxes (see Macdonald, 1976 and Macdonald, et al., 1980, for a different viewpoint).

100 Instinctual basis for learned behavior. For interesting discussions of this important concept, see Eibl-Eibesfeldt (1963, 1970), Hailman (1967), Klopfer (1967), Leyhausen (1973, 1979), Wilson (1975), and Brown (1975).

104 Most researchers have described the red fox as an animal that scatter hoards its caches (see, among others, Murie, 1936; Scott, 1943; Tembrock, 1957a,b; Kruuk, 1964; Tinbergen, 1965; Errington, 1967; Fox, 1971a; Macdonald, 1976; and Dekker, 1983).

104 B. Skelpkovych (pers. comm.) of Memorial University has recently documented an interesting case of red foxes larder-caching prey (also see Fisher, 1951). He studied red foxes on Baccalieu Island, Newfoundland, where a few foxes exist but face no competition from other predators. Thousands of seabirds nest on the island each year, and during the spring and summer foxes capture or scavenge many seabirds, eating or caching them. These foxes often place several birds within the same caching hole. During the remainder of the year, because few alternative prey exist red foxes subsist largely on berries, carrion, and previously made caches.

104 Desmond Morris (1962) was one of the first to suggest that scattering caches may reduce the amount of food that is lost to pilfering scavengers.

106 Tinbergen et al. (1967) was one of the first to test in field experiments that spacing-out is an effective defense against predation. Croze (1970), Göransson et al. (1975), and Andersson and Wiklund (1978) further experimented with this principle (see Curio, 1976, for a general review).

109 For research into the olfactory sense of birds, see Pumphrey (1948), Bang (1960), Portmann (1961), Stager (1964), Wenzel (1967), Grubb (1972).

150

Page

115   Radio-tracking studies of red foxes—see Sargeant (1972), Ables (1975), Pils and Martin (1978), Macdonald (1980d), and Niewold (1980).

116   Andersson and Krebs (1978) discuss an interesting model, which describes caching "cheaters" (that is, those who do not store food themselves but parasitize the caches made by other individuals) and a mechanism by which this strategy can be maintained within a population.

7   KEEPING THINGS STRAIGHT

118   "Keeping Things Straight" is a summary of a detailed research paper entitled "The Use of Urine Marking in the Scavenging Behavior of the Red Fox *(Vulpes vulpes)*" (see Henry, 1977). The reader should refer to this paper for details of experimental design, statistical analysis of the results of each experiment, and a discussion of the limitations inherent in each experiment. The summary of points presented here does not propose to take the place of the more detailed discussions found in this paper.

118   Hediger (1949), Kleiman (1966), Ewer (1968), Mykytowycz (1970, 1971), Ralls (1971), Eisenberg and Kleiman (1972), Johnson (1973), and Müller-Schwarze and Silverstein (1980) offer useful reviews of scent marking among mammals (see also Gleason and Reynierse, 1969 and Bronson, 1971). For scent marking in specific species, see Beach and Gilmore (1949), Bruce (1959), Marsden and Bronson (1964), Moore (1965), Schultze-Westrum (1965), Dominic (1966), Müller-Velton (1966), Michael and Keverne (1968), Wickler (1968), Hornocker (1969), Mykytowycz and Gambale (1969), Lawick-Goodall (1970), Dixon and Mackintosh (1971), Kleiman (1971), Müller-Schwarze (1971, 1974), and Rasa (1973a).

118   Red foxes show other types of chemocommunication, for example, anal sac secretions (Albone et al., 1977) and secretions from the supracaudal scent gland located on the dorsal surface of the tail (see Winton, 1899; Fox, 1975; and Albone and Flood, 1976).

119   For an interesting experimental study of the patterns by which red foxes urine mark their territories, see Macdonald (1979a, 1980b).

119   Regarding the problem of differentiating urination from urine marking behavior, I used the following criteria: urination implies a simple elimination of waste products. On the other hand, urine marking implies an elimination of waste products, where glandular secretions may or may not be added to the urine, and where some of these substances (but not necessarily all of them) are perceived and responded to by conspecifics (adapted from Schenkel, 1966).

119   For a classification of urine marking postures observed in Canidae, see Kleiman (1966), Sprague and Anikso (1973), Henry (1977), and Bekoff and Wells (1985).

119   By way of comparison with other canid species, a description of communication by means of urine marks is available for the wolf (see Peters and Mech, 1975) and the coyote (see Bekoff and Wells, 1985). See also Beach and Gilmore, (1949), Kleiman, (1966), and Sprague and Anikso, 1973.

119   Urine marks serve as a "social scent post" in a number of canid species. See Tinbergen (1965), Burrows (1968), Fox (1971a, 1975), Peters and Mech (1975), Macdonald (1979a), and Bekoff and Wells (1985).

120   Other researchers have commented on the tendency of the red fox and certain other fox species to urine mark food remnants (see, for example, Tembrock, 1957a,b; Egoscue, 1962; Tinbergen, 1965; Burrows, 1968; and Korytin et al., 1969a,b).

128   Heimberger (1959) offers observations that suggest that the urine marks of a red fox are perceived by conspecifics for at least 48 hours.

129   A hierarchy of stimuli has been observed to influence the behavior of other carnivores (see, for example, Rasa, 1973b).

130   Quotes taken from F. H. Harrington. (1981) Urine-marking and caching behavior in the wolf. *Behaviour* 76 (3–4): 280–288.

8   NEW FOXES, NEW ICE

131   Quote taken from L. Eiseley. 1969. "The Innocent Fox." *The Unexpected Universe.* Harcourt Brace Jovanovich Inc. New York. (reprinted by permission).

132   The topic of cognitive and emotional experiences in animals is a controversial one presently receiving critical review (see Kawai, 1965; Griffin, 1981, 1984; Beck, 1980, 1982; and Walker, 1983). Donald R. Griffin, through his recent writings (see Griffin, 1977, 1978, 1981, 1982, 1984), is largely responsible for rekindling this historical debate and has contributed valuable insights to it. Basically Professor Griffin maintains that there is a large body of circumstantial evidence to suggest that animals may have nonverbal thoughts and emotions. But he maintains that the behavioristic assumptions predominating much of psychobiological research today excludes hypotheses concerning subjective mental experiences in animals from being seriously considered. Furthermore, he suggests that only when ethologists and psychologists maintain a spirit of inquisitive yet critical openness about these questions will progress be made in answering them. For viewpoints substantially compatible with Griffin's position, see Darwin (1872), Lorenz (1963), Dennett (1983), and Walker (1983). For viewpoints disputing the idea that we can ever know if animals have cognitive or emotional experiences, see Jaynes (1969), Sebeok et al. (1980, 1981), Harnad (1982), and Roitblat et al. (1983).

138   For the term "innocent killers," I am obviously indebted to Hugo van Lawick and Jane Goodall (see Lawick-Goodall, 1970).

Bibliography

Ables, E. D. 1969a. Home range studies of red foxes *(Vulpes vulpes)*. Journal of Mammalogy 50: 108–120.

———. 1969b. Activity studies of red foxes in southern Wisconsin. Journal of Wildlife Management 33: 145–153.

———. 1975. Ecology of the red fox in America. *In:* M. W. Fox (ed.) *The Wild Canids.* Van Nostrand Reinhold Co. New York.

Albone, E. S. and P. F. Flood. 1976. The supracaudal scent gland of the red fox *(Vulpes vulpes).* Journal of Chemical Ecology 2: 167–175.

———, P. E. Gosden, G. C. Ware, D. W. Macdonald, and N. G. Hough. 1977. Bacterial actions and chemical signaling in the red fox, *Vulpes vulpes,* and other mammals. Proceedings of the American Chemical Society Symposium 67: 78–91.

Aldrich, J. W. 1963. Life areas of North America. Journal of Wildlife Management 27: 147–156.

Allen, D. 1979. *Wolves of Minong: Their Vital Role in a Wild Community.* Houghton-Mifflin Co. Boston.

Allison, L. M. 1971. Activity and behaviour of red foxes in central Alaska. M. Sc. Thesis. (Unpublished) Univ. of Toronto. 92 p.

Andersson, M. and J. Krebs. 1978. On the evolution of hoarding behaviour. Animal Behaviour 26: 707–711.

——— and C. G. Wiklund. 1978. Clumping versus spacing out: Experiments on nest predation in fieldfares *(Turdus pilaris).* Animal Behaviour 26: 1207–1212.

Bacon, P. J. and D. W. Macdonald. 1980. To control rabies: Vaccinate foxes. New Scientist (28 August 1980). pp. 640–645.

Bang, B. G. 1960. Anatomical evidence for olfactory function in some species of birds. Nature (London) 188: 547–548.

Barash, D. P. 1974. Neighbor recognition in two "solitary" carnivores: the raccoon *(Procyon lotor)* and the red fox *(Vulpes fulva).* Science (New York) 185: 794–796.

———. 1975. Behavior as evolutionary strategy. Science (New York) 190: 1084–1085.

Beach, F. A. and R. W. Gilmore. 1949. Response of male dogs to urine from females in heat. Journal of Mammalogy 30: 391–392.

Beck, B. B. 1980. *Animal Tool Behavior.* Garland STPM Press, New York.

———. 1982. Chimpocentrism: bias in cognitive ethology. Journal of Human Evolution 11: 3–17.

Bekoff, M. 1972. The development of social interaction, play, and metacommunication in mammals: an ethological perspective. Quarterly Review of Biology 47: 412–434.

————. 1974. Social play and play-soliciting by infant canids. American Zoologist 14: 323–340.

————. 1975. Social behavior and ecology of African Canidae: a review. *In:* M. W. Fox (ed.) *The Wild Canids.* Van Nostrand Reinhold Co. New York.

————. (ed.) 1978. *Coyotes: Biology, Behavior, and Management.* Academic Press. New York.

———— and M. C. Wells. 1980. Social ecology of coyotes. Scientific American 242: 130–148.

————, J. Diamond, and J. B. Mitton. 1981. Life-history patterns and sociality in canids: Body size, reproduction, and behavior. Oecologia 50: 386–390.

————, M. Tyrrell, V. E. Lipetz, and R. Jamieson. 1981. Fighting patterns in young coyotes: Initiation, escalation, and assessment. Aggressive Behavior 7: 225–244.

———— and M. C. Wells. 1982. Behavioral ecology of coyotes: Social organization, rearing patterns, space use, and resource defense. Zeitschrift für Tierpsychologie 60: 281–305.

————, T. J. Daniels, and J. L. Gittleman. 1984. Life history patterns and the comparative social ecology of carnivores. Annual Review of Ecology and Systematics 15: 191–232.

———— and M. C. Wells. 1985. Social ecology and behavior of coyotes. Advances in the Study of Behavior 16: 251–338.

Bindra, D. 1948a. What makes a rat hoard? Journal of Comparative Physiological Psychology 41: 397–402.

————. 1948b. The nature of motivation for hoarding food. Journal of Comparative Physiological Psychology 41: 411–418.

Bögel, K., H. Moegle, F. Knorpp, A. Arata, K. Dietz, and P. Diethelm. 1976. Characteristics of the spread of a wildlife rabies epidemic in Europe. Bull. World Health Organization 54: 433–447.

Bonner, J. T. 1980. *The Evolution of Culture in Animals.* Princeton University Press. Princeton, New Jersey.

Bowen, W. D. 1981. Variation in coyote social organization: The influence of prey size. Canadian Journal of Zoology 59: 639–652.

Bronson, F. H. 1971. Rodent pheromones. Biology of Reproduction 4: 344–357.

Brown, J. L. 1975. *The Evolution of Behavior.* Norton. New York.

Bruce, H. M. 1959. An exteroceptive block to pregnancy in the mouse. Nature (London) 184: 105.

Buddenbrock, W. von. 1959. *The Senses.* University of Michigan Press. Ann Arbor.

Bueler, L. E. 1973. *Wild Canids of the World.* Stein and Day Publishers. New York.

Burrows, R. 1968. *Wild Fox.* David Charles Ltd. Newton Abbot, England.

Burt, W. H. and R. Grossenheider. 1974. *A Field Guide to the Mammals.* The Peterson Field Guide Series. Houghton Mifflin Co. Boston.

Carbyn, L. N. 1974. Wolf predation and behavioural interaction with elk and other ungulates in an area of high prey diversity. Canadian Wildlife Service. 233p. Mimeo.

————. (ed.) 1981. Wolves in Canada and Alaska: Their status, biology, and management. Canadian Wildlife Service Report Series Number 45. 135 pp.

Chesemore, D. L. 1975. The arctic fox *(Alopex lagopus)* in North America: a review. *In:* M. W. Fox (ed.) *The Wild Canids.* Van Nostrand Reinhold Co. New York.

Churcher, C. S. 1959. The specific status of the New World red fox. Journal of Mammalogy 40: 513–520.

————. 1960. Cranial variation in the North American red fox. Journal of Mammalogy 41: 349–360.

Clutton-Brock, J., G. B. Corbet, and M. Hills. 1976. Review of the family Canidae, with a classification by numerical methods. Bulletin of the British Museum (Natural History) Zool. 29: 119–199.

Colbert, E. H. 1980. *Evolution of the Vertebrates.* 3rd ed. John Wiley and Sons. New York.

Coman, B. J. 1973. The diet of red foxes, *Vulpes vulpes* L., in Victoria. Australian Journal of Zoology 21: 391–401.

———— and H. Brunner. 1972. Food habits of the feral house cat in Victoria. Journal of Wildlife Management 36: 848–853.

Conde, B., Nguyen-thi-thu-cuc, F. Vailiant, and P. Schauenberg. 1972. Le regime alimentaire du chat forestier (*F. silvestris* Schr.) en France. Mammalia 36: 112–119.

Cook, D. B. and W. J. Hamilton, Jr. 1944. The ecological relationship of red fox food in eastern New York. Ecology 25: 91–104.

Cott, H. B. 1940. *Adaptive Colouration in Animals.* Methuen. London.

Crook, J. H. 1965. The adaptive significance of avian social organizations. Symposium of the Zoological Society of London 14: 181–218.

Crowcroft, P. 1957. *The Life of a Shrew.* Max Reinhardt. London.

Croze, H. 1970. Searching image in carrion crows. Zeitschrift für Tierpsychologie Beiheft 5. 85p.

Cullen, E. 1957. Adaptations in the kittiwake to cliff-nesting. Ibis 99: 275–302.

Curio, E. 1975. The functional organization of anti-predator behaviour in the pied flycatcher. A study of avian perception. Animal Behaviour. 23: 1–115.

————. 1976. *The Ethology of Predation.* Springer-Verlag. Berlin.

Darwin, C. 1872. *The Expression of Emotions in Man and Animals.* John Murray. London.

Davis, D. D. 1964. The giant panda: A morphological study of evolutionary mechanisms. Fieldiana: Zoology Memoirs Vol. 3. Chicago Natural History Museum.

Dekker, D. 1976. Mortality rates of red fox pups, and causes of death of adult foxes in central Alberta. Alberta Naturalist 6: 65–67.

————. 1983. Denning and foraging habits of red foxes, *Vulpes vulpes,* and their interactions with coyotes, *Canis latrans,* in central Alberta, 1972–1981. Canadian Field Naturalist 97: 303–306.

Dennett, D. C. 1983. Intentional systems in cognitive ethology: The "Panglossian Paradigm" defended. Behavior and Brain Science 6: 343–390.

Dixon, A. K. and J. H. Mackintosh. 1971. Effects of female urine upon the social behaviour of adult male mice. Animal Behaviour 19: 138–140.

Dobbs, D. G. 1955. Food habits of the Newfoundland red fox. Journal of Mammalogy 36: 291–298.

Dominic, C. J. 1966. Block to pseudopregnancy in mice caused by exposure to male urine. Experientia 22: 534.

Dücker, G. 1957. Farb- und Helligkeitsschen und Instinke bei Viverriden und Feliden. Zoologische Beitraege (Berlin) 3: 25–99.

Eadie, W. R. 1943. Food of the red fox in southern New Hampshire. Journal of Wildlife Management 7: 74–77.

Eaton, R. L. 1969. Hunting behavior of the cheetah. Journal of Wildlife Management 34: 56–67.

————. 1970. The predatory sequence, with emphasis on killing and its ontogeny, in the cheetah (*Acinonyx jubatus* Schreber). Zeitschrift für Tierpsychologie 27: 492–504.

————. 1974. *The Cheetah: Its Behavior and Ecology.* Van Nostrand Reinhold Co. New York.

Eberhard, T. 1954. Food habits of Pennsylvania house cats. Journal of Wildlife Management 18: 284–286.

Edmunds, M. 1974. *Defence in Animals: A Survey of Anti-Predator Defences.* Longman Inc. New York.

Egoscue, H. J. 1962. Ecology and life history of the kit fox in Toole County, Utah. Ecology 43: 481–497.

Ehrlich, P. R. and R. Holm. 1963. *The Process of Evolution.* McGraw-Hill Book Co. San Francisco.

Eibl-Eibesfeldt, I. 1963. Angeborenes und Erworbenes im Verhaltens einiger Säuger. Zeitschrift für Tierpsychologie 20: 705–754.

———. 1970. *Ethology: The Biology of Behavior.* Holt, Rinehart, and Winston Co. New York.

Eisenberg, J. F. 1981. *The Mammalian Radiations: An Analysis of Trends in Evolution, Adaptations and Behavior.* University of Chicago Press. Chicago.

——— and D. G. Kleiman. 1972. Olfactory communication in mammals. Annual Review of Ecology and Systematics 3: 1–32.

——— and P. Leyhausen. 1972. The phylogenesis of predatory behavior in mammals. Zeitschrift für Tierpsychologie 30: 59–93.

English, P. F. and L. J. Bennett. 1942. Red fox food habits study in Pennsylvania. Pennsylvania Game News 12: 6–23.

Englund, J. 1965. Studies of food ecology of the red fox *(Vulpes vulpes)* in Sweden. Viltrevy 3: 377–485.

———. 1970. Some aspects of reproduction and mortality rates in Swedish red foxes *(Vulpes vulpes)*, 1961–63 and 1966–69. Viltrevy 8: 1–82.

———. 1980a. Population dynamics of the red fox *(Vulpes vulpes)* in Sweden. Biogeographica 18: 107–122.

———. 1980b. Dispersal rates of fox cubs tagged in Swedish coniferous forests. Biogeographica 18: 195–207.

Ennion, E. A. R. and N. Tinbergen. 1967. *Tracks.* Oxford University Press. Oxford. England.

Errington, P. L. 1935. Food habits of mid-west foxes. Journal of Mammalogy 16: 192–200.

———. 1967. *Of Predation and Life.* Iowa State University Press. Ames, Iowa.

Estes, R. D. 1966. Behaviour and life history of the wildebeest. Nature (London) 212: 999–1000.

———. 1967. Predators and scavengers. Parts I and II. Natural History 76: 21–29 and 38–47.

———. 1974. Social Organization of the African Bovidae. *In:* V. Geist and F. Walther (eds.) *The Behaviour of Ungulates and Its Relation to Management.* International Union for the Conservation of Nature, new series, no. 24. Morges, Switerland.

——— and J. Goddard. 1967. Prey selection and hunting behavior of the African wild dog. Journal of Wildlife Management 31: 52–70.

Ewer, R. F. 1968. *Ethology of Mammals.* Plenum Press. New York.

———. 1969. Some observations on the killing and eating of prey by two dasyurid marsupials. Zeitschrift für Tierpsychologie 26: 23–38.

———. 1973. *The Carnivores.* Weidenfeld and Nicholson. London.

——— and C. Wemmer. 1974. The behaviour in captivity of the African civet, *Civettictis civetta* (Schreber). Zeitschrift für Tierpsychologie 34: 360–394.

Fairley, J. S. 1970. The food, reproduction, form, growth, and development of the fox *Vulpes vulpes* in north-east Ireland. Proceedings of the Royal Irish Academy 69: 103–137.

Fentress, J. C. 1968. Interrupted ongoing behavior in two species of vole. Parts I and II. Animal Behavior 16: 135–167.

Findley, J. S. 1956. Comments on the winter food of red foxes in eastern South Dakota. Journal of Wildlife Management 6: 221–224.

Fiorone, F. 1970. *The Encyclopedia of Dogs: The Canine Breeds.* T. Y. Crowell Co. New York.

Fisher, H. I. 1951. Notes on the red fox *(Vulpes fulva)* in Missouri. Journal of Mammalogy 32: 296–300.

Fishman, J. Y. 1971. Taste responses in the red fox. Physiological Zoology 44: 171–176.

Fox, M. W. 1969a. Ontogeny of prey-killing in Canidae. Behaviour 35: 259–272.

———. 1969b. The anatomy of aggression and its ritualization in Canidae: a developmental and comparative study. Behaviour 35: 242–258.

———. 1970. A comparative study of the development of facial expressions in canids: Wolf, coyote, and foxes. Behaviour 36: 49–73.

———. 1971a. *The Behaviour of Wolves, Dogs, and Related Canids.* Jonathan Cape. London.

———. 1971b. Socio-infantile and socio-sexual signals in canids: A comparative and ontogenentic study. Zeitschrift für Tierpsychologie 28: 185–210.

———. 1975. *The Wild Canids: Their Systematics, Behavioral Ecology, and Evolution.* Van Nostrand Reinhold Co. New York.

Frisch, K. von and O. von Frisch. 1974. *Animal Architecture.* Harcourt Brace and Jovanovich Inc. New York.

Frisch, O. von and H. von Frisch. 1971. Beobochtungen bei der Handaufzucht und spaeteren Aussetzung einer Fuchsfaeke *Vulpes vulpes.* Zeitschrift für Tierpsychologie 28: 534–541.

Gambaryan, P. P. 1974. *How Animals Run.* Keter Publishing Ltd. Jerusalem.

Gauthier-Pilters, H. 1962. Beobachtungen an Feneks *(Fennecus zerda,* Zimm.). Zeitschrift für Tierpsychologie 19: 440–464.

———. 1967. The fennec. African Wildlife 21: 117–125.

Geiger, G., J. Brömel, and K. H. Habermehl. 1977. Konkordanz verschiedener Methoden der altersbestimmung beim Rotfuchs *(Vulpes vulpes* L. 1758). Zeitschrift für Jagdwissenschaft 23: 57–64.

Geist, V. 1972. An ecological and behavioral explanation of mammalian characteristics, and their implication to therapsid evolution. Zeitschrift für Saugetierkunde 37: 1–15.

———. 1975. *Mountain Sheep and Man in the Northern Wilds.* Cornell University Press. New York.

Gimbarzevsky, P. 1973. Prince Albert National Park: Natural resources (land, vegetation, and open water). Canadian Dept. of the Environment, Canadian Forestry Service, Forest Management Institute. 81p.

Girvin, D. and W. J. Beecher. 1967. Kit Fox: A Conservation Case Study. Journal Films, Inc. Chicago.

Gittleman, J. L. and P. H. Harvey. 1982. Carnivore home-range size, metabolic needs and ecology. Behavioral Ecology and Sociobiology 10: 57–63.

Gleason, K. K. and J. H. Reynierse. 1969. The behavioral significance of pheromones in vertebrates. Psychological Bulletin 71: 58–73.

Goethe, F. 1950. Das Verhalten des Musteliden. Handbuch der Zoologie VIII, Part 10 (37): 1–80.

Gould, S. J. 1979. This view of life: Mickey Mouse meets Konrad Lorenz. Natural History 88: 30–36.

Göransson, G., J. Karlsson, S. G. Nilsson, and S. Ulfstra.id. 1975. Predation on bird's nests in relation to antipredator aggression and nest density: an experimental approach. Oikos 26: 117–120.

Goszczynski, J. 1974. Studies on the food of foxes. Acta Theriologica 19: 1–18.

Gray, D. R. 1970. The killing of a bull muskox by a single wolf. Arctic 23: 197–199.

Griffin, D. R. 1977. Anthropomorphism. BioScience 18: 445–446.

———. 1978. Prospects for a cognitive ethology. Behavioral Brain Science 1: 527–538.

———. 1981. *The Question of Animal Awareness*. 2nd ed. Rockefeller University Press and William Kaufmann Inc. Los Altos, California.

———. (ed.) 1982. *Animal Mind–Human Mind*. Springer-Verlag. New York.

———. 1984. *Animal Thinking*. Harvard University Press. Cambridge, Massachusetts.

Grubb, T. C. Jr. 1972. Smell and foraging in shearwaters and petrils. Nature (London) 237: 404–405.

Gustavsson, I. and C. O. Sundt. 1967. Chromosome complex of the family of Canidae. Hereditas 54: 249–254.

Hailman, J. P. The ontogeny of an instinct. Behavioral Supplement 15: 1–159.

Haines, R. W. 1932. The laws of muscle and tendon growth. Journal of Anatomy (London) 66: 578-585.

Hall, B. K. 1975. Evolutionary consequences of skeletal differentiation. American Zoologist 15: 329–350.

Hamilton, W. D. 1964. The genetical evolution of social behavior. Journal of Theoretical Biology 7: 1–52.

———. 1971. Geometry for the selfish herd. Journal of Theoretical Biology 31: 295–311.

———. 1972. Altruism and related phenomena, mainly in social insects. Annual Review of Ecology and Systematics 3: 193–232.

Hamilton, W. J. Jr. 1935. Notes on food of red foxes in New York and New England. Journal of Mammalogy 16: 16–21.

Harrington, F. H. 1981. Urine-marking and caching behavior in the wolf. Behaviour 76: 280–288.

———, L. D. Mech, and S. Fritts. 1983. Pack size and wolf pup survival: Their relationship under varying ecological conditions. Behavioral Ecology and Sociobiology 13: 19–26.

Harris, S. 1977. Distribution, habitat utilization and age structure of a suburban fox *(Vulpes vulpes)* population. Mammalogy Review 7: 25–39.

———. 1978. Age determination in the red fox *(Vulpes vulpes)*—an evaluation of technique efficiency as applied to a sample of suburban foxes. Journal of Zoology (London) 184: 97–117.

Hatfield, D. M. 1939. Winter food of foxes in Minnesota. Journal of Mammalogy 20: 202–206.

Hediger, H. 1949. Säugetier - Territorien und ihre Markierung. Bijdragen tot de Dierkunde 28: 172–184.

Heidemann, G. and G. Vauk. 1970. Zur Nahrungsokologie "wildernder" Hauskatzen. Zeitschrift für Saugetierkunde 35: 185–190.

———. 1973. Weitere Untersuchungen zur Nahrungsokologie "wildernder" Hauskatzen *(Felis sylvestris catus* Linne, 1758). Zeitschrift für Saugetierkunde 38: 216–224.

Heimburger, N. 1959. Das Markierungsverhalten einiger Caniden. Zeitschrift für Saugetierkunde 16: 104–113.

Henry, J. D. 1976. Adaptive strategies in the behaviour of the red fox, *Vulpes vulpes* L. Ph.D. Thesis. (Unpublished) University of Calgary. 259 pp.

———. 1977. The use of urine marking in the scavenging behavior of the red fox *(Vulpes vulpes)*. Behaviour 61: 82–106.

———. 1980a. Fox hunting. Natural History 89: 60–69.

————. 1980b. The urine marking behavior and movement patterns of red foxes *(Vulpes vulpes)* during a breeding and post-breeding period. In: D. Müller-Schwarze and R. M. Silverstein (eds.) *Chemical Signals: Vertebrates and Aquatic Invertebrates.* pp. 11–27. Plenum Press. New York.

————. 1985. The little foxes. Natural History 94: 46–57.

Henry, J. D. and S. M. Herrero. 1974. Social play in the American black bear: its similarity to canid social play and an examination of its identifying characteristics. American Zoologist 14: 371–390.

Henshaw, R. E., L. S. Underwood, and T. M. Casey. 1972. Peripheral thermoregulation in two Arctic canines. Science (New York) 175: 988–990.

Hersteinsson, P. and D. W. Macdonald. 1982. Some comparisons between red and arctic foxes, *Vulpes vulpes* and *Alopex lagopus,* as revealed by radio tracking. Symp. Zool. Soc. Lond. No. 49: 259–289.

Hewson, R. and H. Kolb. 1973. Changes in numbers and distribution of foxes *(Vulpes vulpes)* killed in Scotland from 1948–70. Journal of Zoology (London). 171: 285–292.

Hildebrand, M. 1952a. An analysis of body proportions in the Canidae. American Journal of Anatomy 90: 217–256.

————. 1952b. The integument in Canidae. Journal of Mammalogy 33: 419–428.

————. 1954. Comparative morphology of the body skeleton in recent Canidae. University of California, Publications in Zoology 52: 399–470.

————. 1959. Motions of the running cheetah and horse. Journal of Mammalogy. 40: 481–495.

————. 1961. Further studies on the locomotion of the cheetah. Journal of Mammalogy 42: 84–91.

————. 1974. *Analysis of Vertebrate Structure.* John Wiley and Sons. New York.

Hocking, B. and B. L. Mitchell. 1961. Owl vision. Ibis 103: 284–288.

Hockman, J. G. and J. A. Chapman. 1983. Comparative feeding habits of red foxes *(Vulpes vulpes)* and gray foxes *(Urocyton cinereoargenteus)* in Maryland. American Midland Naturalist. 110: 276–285.

Horn, H. S. 1975. Tree architecture. Scientific American 232: 90–98.

Hornocker, M. G. 1969. Winter territoriality in mountain lions. Journal of Wildlife Management 33: 457–464.

————. 1970. An analysis of mountain lion predation upon mule deer and elk in the Idaho Primitive area. Wildlife Monographs 21: 1–39.

Howell, A. B. 1944. *Speed in Animals.* University of Chicago Press. Chicago.

Humphries, D. A. and P. M. Drivers. 1967. Erratic display against predators. Science (New York) 156: 1767–1768.

———— and ————. 1971. Protean defence by prey animals. Oecologia 5: 285–302.

Hunt, G. L. and M. W. Hunt. 1975. Reproductive ecology of the western gull: the importance of nest spacing. Auk 92: 270–279.

Hurrell, H. G. 1962. Foxes. Sunday Times (London) Publications on Animals of Britain No. 9.

Huxley, T. H. 1880. On the cranial and dental characters of the Canidae. Proceedings of the Zoological Society (London). 238–288.

Isley, T. E. and L. W. Gysel. 1975. Sound source location by the red fox. Journal of Mammalogy 56: 397–404.

Jaynes, J. 1969. The historical origins of 'Ethology' and 'Comparative Psychology'. Animal Behaviour 17: 601–606.

Jeglum, J. K. 1972. Boreal forest wetlands near Candle Lake, central Saskatchewan. I. Vegetation. Musk-Ox 11: 41–58.

————. 1973. Boreal forest wetlands near Candle Lake, central Saskatchewan. II. Relationships of vegetation to major environmental gradients. Musk-Ox 12: 32–48.

Jensen, B. and D. M. Sequeira. 1978. The diet of the red fox (*Vulpes vulpes* L.) in Denmark. Danish Review of Game Biology 10: 1–16.

Johnson, R. P. 1973. Scent marking in mammals. Animal Behaviour 21: 521–535.

Johnson, W. J. 1970. Food habits of the red fox in Isle Royale National Park, Lake Superior. American Midland Naturalist 84: 568–572.

Johnston, D. H. and M. Beauregard. 1969. Rabies epidemiology in Ontario. Bulletin of the Wildlife Disease Association 5: 357–370.

Jones, D. M. and J. B. Theberge. 1982. Summer home range and habitat utilization of the red fox (*Vulpes vulpes*) in a tundra habitat, northwest British Columbia. Canadian Journal of Zoology 60: 807–812.

———— and ————. 1983. Variation in red fox, *Vulpes vulpes*, summer diets in northwest British Columbia and southwest Yukon. Canadian Field Naturalist 97: 311–314.

Jorgenson, J. W., M. Novotny, M. Carmack, G. B. Copland, S. R. Wilson, S. Katona, and W. K. Whitten. 1978. Chemical scent constituents in the urine of the red fox (*Vulpes vulpes* L.) during the winter season. Science (New York) 199: 796–798.

Karlson, P. and M. Lüscher. 1959. Pheromones: a new term for a class of biologically active substances. Nature (London) 183: 55–56.

Karpuleon, F. 1958. Food habits of Wisconsin foxes. Journal of Mammalogy 39: 591–593.

Kaufman, D. K. 1974. Adaptive coloration in *Peromyscus polionotus:* experimental selection by owls. Journal of Mammalogy 55: 271–283.

Kavanau, J. L. 1967. Behavior of captive white footed mice. Science (New York) 155: 1623–1639.

————. 1971. Locomotion and activity phasing of some medium-size mammals. Journal of Mammalogy 52: 386–403.

———— and J. Ramos. 1972. Twilights and onset and cessation of carnivore activity. Journal of Wildlife Management 36: 653–657.

Kawai, M. 1965. Newly acquired pre-cultural behavior of the natural troop of Japanese monkeys on Koshima Islet. Primates 6: 1–30.

Keith, L. B. 1963. *Wildlife's Ten-Year Cycle.* University of Wisconsin Press. Madison.

————. 1974. Some features of population dynamics in mammals. Proceedings of the International Congress of Game Biologists 11: 17–58.

Kleiman, D. G. 1966. Scent Marking in the Canidae. Symposium of the Zoological Society (London) 18: 167–177.

————. 1967. Some aspects of social behavior in the Canidae. American Zoologist 7: 365–372.

Kleiman, D. G. 1977. Monogamy in mammals. Quarterly Review of Biology 52: 39–69.

———— and J. F. Eisenberg. 1973. Comparisons of canid and felid social systems from an evolutionary perspective. Animal Behaviour 21: 637–659.

Klopfer, P. H. 1969. Instincts and chromosomes: What is an "innate" act? American Naturalist. 103: 556–560.

Koenig, L. 1970. Zur Fortpflanzung und Jugendentwicklung des Wüstenfuchses (*Fennecus zerda*, Zimm. 1780). Zeitschrift für Tierpsychologie 27: 205–246.

Kolb, H. H. 1984. Factors affecting the movements of dog foxes in Edinburgh. Journal of Applied Ecology 21: 161–173.

Konishi, M. 1973. How the owl tracks its prey. American Scientist 61: 414–424.

Korschgen, L. J. 1959. Food habits of the red fox in Missouri. Journal of Wildlife Management 23: 168–176.

Korytin, S. A. and M. D. Azbukina. 1969. Sezonnaya i polovaya izmenchivost' reaktsii pestsov na zapakhi. Sb. Trud. vses. nauchno-issled. Inst. Zhivotnogo Syr'ya PhsHniny 22: 223–234.

—— and N. N. Solomin. 1969. Materialy po etiologii psovykh. Sb. Trud. vses. nauchno-issled. Inst. Zhivotnogo Syr'ya PusHniny 22: 235–270.

Kruuk, H. 1964. Predators and anti-predator behaviours of the black-headed gull (Larus ridibundus L.) Behaviour Supplement 11: 1–130.

——. 1972. The Spotted Hyena. University of Chicago Press. Chicago.

——. 1976. The biological function of gulls' attraction towards predators. Animal Behaviour 24: 146–153.

—— and M. Turner. 1967. Comparative notes on predation by lions, leopards, cheetahs, and wild dogs in the Serengeti area, East Africa. Mammalia 31: 1–27.

Kühme, W. 1965a. Communal food distribution and division of labour in African hunting dogs. Nature (London) 205: 443–444.

——. 1965b. Freilandstudien zur Sociologie des Hyänenhundes Lycaon pictus lupinus Thomas (1902). Zeitschrift für Tierpsychologie 22: 495–441.

Kuyt, E. 1972. Food habits and ecology of wolves on barren-ground caribou range in the Northwest Territories. Canadian Wildlife Service Report Series No. 21. 36p.

Langguth, A. 1969. Die südamerikanischen Canidae unter besonderer Berücksichtingung des Mähnenwolfes, Chrysocyon brachyurus, Illiger. Zeitschrift für Wissenschaftliche Zoologie 179: 1–188.

——. 1970. Una nueva clasificación de los canidos sudamericanos. Acta de IV Congreso de Latina Zoologia 1: 129–143.

——. 1975. Ecology and evolution in the South American canids. In: M. W. Fox (ed.) The Wild Canids. Van Nostrand Reinhold Co. New York.

Latham, R. M. 1950. The food of predaceous animals in northeastern United States. Pennsylvania Game Commission Bulletin 69 p.

Lawick, H. van. 1974. Solo: The Story of an African Hunting Dog. Houghton Mifflin Co. Boston.

—— and J. van Lawick-Goodall. 1970. Innocent Killers. Collins Ltd. London.

Layne, J. N. and W. H. McKeon. 1956a. Notes on the development of the red fox fetus. New York Fish and Game Journal. 3: 120–128.

—— and ——. 1956b. Notes on red fox and gray fox den sites in New York. New York Fish and Game Journal 3: 248–249.

Lever, R. A. 1959. Diet of the fox since myxomatosis. Journal of Animal Ecology 28: 359–375.

Leyhausen, P. 1956. Verhaltensstudien der Katzen. Zeitschrift für Tierpsychologie Beiheft 2. 120 p.

——. 1965a. The communal organization of solitary animals. Symposium of the Zoological Society (London). 14: 249–263.

——. 1965b. Über die Funktion der relativen Stimmungshierarchie. Zeitschrift für Tierpsychologie 22: 412–494.

——. 1973. Verhaltensstudien der Katzen. Zeitschrift für Tierpsychologie Beiheft 2. 3rd. ed. 133 p.

——. 1979. Cat Behavior. Garland STPM Press. New York.

Lindström, E. 1980. The red fox in a small game community of the south taiga region in Sweden. Biogeographica 18: 177–184.

——. 1983. Condition and growth of red foxes (Vulpes vulpes) in relation to food supply. Journal of Zoology (London). 199: 177–222.

Lloyd, H. G. 1980a. *The Red Fox*. Batsford Co. Ltd. London.

———. 1980b. Habitat requirements of the red fox. Biogeographica 18: 7–26.

Lockie, J. D. 1956. After myxomatosis. Scottish Agriculture 36: 65–69.

———. 1959. The estimation of the food of foxes. Journal of Wildlife Management 23: 224–227.

Lorenz, K. Z. 1931. Beiträge zur Ethologie sozialer Corviden. Journal für Ornithologie 79: 67–120.

———. 1935. Der Kumpan in der Umwelt des Vogels. Journal of Ornithology 83: 137–213 and 289–413.

———. 1943. Die angeborenen Formen möglicher Erfahrung. Zeitschrift für Tierpsychologie 5: 235–409.

———. 1971. Do animals undergo subjective experience? *In: Studies in Animal and Human Behavior*. Volume II. Harvard University Press. Cambridge, Massachusetts.

Loškariev, G. A. 1970. Pitanije kavkazkoj lisicy *(Vulpes vulpes caucasica)* v predgoriah Sovernogo Kaukaza. Zoologiceskij Zhurnal 49: 903–907.

Macdonald, D. W. 1976. Food caching by red foxes and some other carnivores. Zeitschrift für psychology 42: 170–185.

———. 1977. On food preference in the red fox. Mammal Review 7: 7–23.

———. 1979a. Some observations and field experiments on the urine marking behavior of the red fox, *Vulpes vulpes* L. Zeitschrift für Tierpsychologie 51: 1-22.

———. 1979b. 'Helpers' in fox society. Nature (London) 282: 69–71.

———. 1980a. The red fox, *Vulpes vulpes*, as a predator upon earthworms, *Lumbricus terrestris*. Zietschrift für Tierpsychologie 52: 171–200.

———. 1980b. Patterns of scent marking with urine and faeces among carnivore communities. Symposium of the Zoological Society (London). 45: 107–139.

———. 1980c. *Rabies and Wildlife: A Biologist's Perspective*. Oxford University Press. Oxford, England.

———. 1980d. Social factors affecting reproduction amongst red foxes *(Vulpes vulpes)*. Biogeographica 18: 123–176.

———. 1983. The ecology of carnivore social behaviour. Nature (London) 301: 379–384.

———, L. Boitani, and P. Barrasso. 1980. Foxes, wolves and conservation in the Abruzzo mountains. Biogeographica 18: 223–236.

——— and P. D. Moehlman. 1982. Cooperation, altruism, and restraint in the reproduction of carnivores. Perspectives in Ethology 5: 433–467.

MacGregor, A. E. 1942. Late fall and winter food of foxes in central Massachusetts. Journal of Wildlife Management 6: 221–224.

Malcolm. J. R. and H. van Lawick. 1975. Notes on wild dogs *(Lycaon pictus)* hunting zebras. Mammalia 39: 231–240.

Marler, P. 1955. Characteristics of some animal calls. Nature (London) 176: 6–8.

Marsden, H. M. and F. H. Bronson. 1964. Estrous synchrony in mice: alteration by exposure to male urine. Science (New York) 144: 3625.

Marten, G. G. 1973. Time patterns of *Peromyscus* activity and their correlations with weather. Journal of Mammalogy 54: 168–169.

Martin, A. C., H. S. Zim, and A. L. Nelson. 1951. *American Wildlife and Plants: A Guide to Wildlife Food Habits*. Dover Publ. New York.

Matthew, W. D. 1930. The phylogeny of dogs. Journal of Mammalogy 11: 117–138.

McCord, C. M. 1974. Selection of winter habitat by bobcats *(Lynx rufus)* on the Quabbin Reservation, Massachusetts. Journal of Mammalogy 55: 428–437.

McMahon, P. 1975. The victorious coyote. Natural History 84: 44–50.

McMahon, T. 1973. Size and shape in biology. Science (New York) 179: 1201–1204.

162

McMurray, F. B. and C. C. Sperry. 1941. Food of feral house cats in Oklahoma. Journal of Mammalogy 22: 185–190.

Mech, L. D. 1966. The wolves of Isle Royale. Fauna of National Parks. U. S. Fauna Series 7.

———. 1970. *The Wolf*. Natural History Press. New York.

Meddis, R. 1975. On the function of sleep. Animal Behaviour 23: 676–691.

Merwe, N. J. van der. 1953. The jackal. Fauna and Flora 4: 4–80.

Messier, F. and C. Barrette. 1982. The social system of the coyote *(Canis latrans)* in a forested habitat. Canadian Journal of Zoology. 60: 1743–1753.

Metzgar, L. H. 1967. An experimental comparison of screech owl predation on resident and transient white-footed mice *(Peromyscus leucopus)*. Journal of Mammalogy 48: 387–391.

Michael, R. P. and E. B. Keverne. 1968. Pheromones in the communication of sexual status in primates. Nature (London) 218: 746–749.

Mivart, S. G. 1890. *Dogs, Jackals, Wolves, and Foxes: a Monograph of the Canidae*. London.

Moehlman, P. D. 1979. Jackal helpers and pup survival. Nature (London) 277: 382–383.

Montgomery, G. G. 1974. Communication in red fox dyads: a computer simulation study. Smithsonian Contribution to Zoology 187: 1–30.

Moore, R. E. 1965. Olfactory discrimination as an isolating mechanism between *Peromyscus maniculatus* and *Peromyscus polionotus*. American Midland Naturalist. 73: 85–100.

Morrell, F. K. 1972. Visual system's view of acoustic space. Nature (London) 238: 44–46.

Morris, D. 1962. The behaviour of the green acouchi *(Myoprocta pratti)* with special reference to scatter hoarding. Proceedings of the Zoological Society (London) 139: 701–732.

Müller-Schwarze, D. 1971. Pheromones in black-tailed deer *(Odocoileus hemionus columbianus)*. Animal Behaviour 19: 141–152.

——— and C. Müller-Schwarze. 1971. Response of chipmunks to aerial predators. Journal of Mammalogy 52: 456–457.

———, ———, A. G. Singer, and R. M. Silverstein. 1974. Mammalian pheromone: identification of the active component in the subauricular scent of the male pronghorn. Science (New York) 183: 860–862.

Müller-Schwarze, D. and R. M. Silverstein. (eds.) 1980. *Chemical Signals: Vertebrates and Aquatic Invertebrates*. Plenum Press. New York.

Müller-Velten, H. 1966. Uber den Angstgeruch bei der Hausmaus (Mus musculus L.). Zeitschrift für Vergleichende Physiologie 52: 401–429.

Murie, A. 1936. Following fox trails. Miscellaneous Publications in Zoology of the University of Michigan 32: 1–45.

———. 1940. Ecology of the coyote in the Yellowstone. U.S.D.I. Fauna Series 4. National Park Service, Washington.

———. 1944. The wolves of Mount McKinley. U.S. National Parks Service Fauna Series 5. Washington.

———. 1961. *A Naturalist in Alaska*. Doubleday and Co. Inc. New York.

Mykytowycz, R. 1970. The role of skin glands in mammalian communication. In; J. W. Johnston, Jr., D. G. Moulton, and A. Turk (eds.) *Advances in Chemoreception. I: Communication by Chemical Signals*. Appleton, Century, Crofts Co. New York.

———. 1971. The behavioral role of the mammalian skin glands. Naturwissenschaffen 59: 133–139.

——— and S. Gambale. 1969. The distribution of dung-hills and the behavior of free-living rabbits on them. Forma et Functio 1: 333–349.

Niewold, F. J. J. 1980. Aspects of the social structure of red fox populations: A summary. Biogeographica 18: 185–193.

Norman, F. I. 1971. Predation by the fox (*Vulpes vulpes* L.) on colonies of the short-tailed shearwater (*Puffinus tenuirostris* Temminck) in Victoria, Australia. Journal of Applied Ecology 8: 21–32.

Orians, G. H. 1969. On the evolution of mating systems in birds and mammals. American Naturalist 103: 589–603.

Österholm, H. 1964. The significance of distance receptors in the feeding behavior of the fox (*Vulpes vulpes* L.) Acta Zoologica Fennica 106: 1–31.

Owens, D. D. and M. J. Owens. 1984. Helping behavior in brown hyenas. Nature (London) 308: 843–845.

Ozoga, J. J. and E. M. Harger. 1966. Winter activities and feeding habits of northern Michigan coyotes. Journal of Wildlife Management 30: 809–818.

Palm, P. O. 1970. Rodravens naringsekologi. Zoologisk Revy 32: 43–46.

Parmalee, P. W. 1953. Food habits of the feral house cat in east-central Texas. Journal of Wildlife Management 17: 375–376.

Peters, R. and L. D. Mech. 1975. Scent-marking in wolves. American Scientist 63: 628–637.

Peterson, F. A., W. C. Heaton, and S. D. Wruble. 1969. Levels of auditory response in fissiped carnivores. Journal of Mammalogy 50: 566–578.

Peterson, R. O. 1977. Wolf ecology and prey relationships on Isle Royale. National Park Service Monograph Series No. 11. 210 pp.

Phillips, R. L. 1970. Age ratios of Iowa foxes. Journal of Wildlife Management 34: 52–56.

———, R. D. Andrews, G. L Storm, and R. A. Bishop. 1972. Dispersal and mortality of red foxes. Journal of Wildlife Management 36: 237–248.

Pils, C. M. and M. A. Martin. 1977. Predator control: A case against red fox reduction in Wisconsin. In: R. L. Phillips and C. J. Jonkel (eds.) Proceedings of the 1975 Predator Symposium. Montana Forest and Conservation Experimental Station, School of Forestry, University of Montana, Missoula.

——— and ———. 1978. Population dynamics, predator-prey relationships and management of the red fox in Wisconsin. Technical Bulletin Number 105. Department of Natural Resources, Madison, Wisconsin. 56 p.

Pimlott, D. H. 1967. Wolf predation and ungulate populations. American Zoologist 7: 267–278.

———, J. A. Shannon, and G. B. Kolenosky. 1969. The ecology of the timber wolf in Algonquin Provincial Park. Ontario Dept. of Lands and Forests Research Report No 87.

Pocock, R. I. 1914a. On the facial vibrissae of mammalia. Proceedings of the Zoological Society (London). 1914: 889–912.

———. 1914b. On the feet and other external features of the Canidae and Ursidae. Proceedings of the Zoological Society (London). 1914: 913–941.

Portmann, A. 1961. Sensory organs: skin, taste, and olfaction. In: A. J. Marshall (ed.) *Biology and Comparative Physiology of Birds*. Academic Press. New York.

Powell, G. V. N. 1974. Experimental analysis of the social value of flocking by starlings (*Sturnus vulgaris*) in relation to predation and foraging. Animal Behaviour 22: 501–505.

Preston, E. M. 1975. Home range defense in the red fox, *Vulpes vulpes* L. Journal of Mammalogy 56: 645–652.

Pumphrey, R. J. 1948. The sense organs of birds. Ibis 90: 171–199.

Radinsky, L. B. 1973. Evolution of the canid brain. Brain, Behaviour and Evolution 7: 169–202.

164

————. 1981. Evolution of skull shape in carnivores. 1. Representative modern carnivores. Biological Journal of the Linnean Society 15: 369–388.

————. 1982. Evolution of skull shape in carnivores. 3. The origin and early radiation of the modern carnivore families. Paleobiology 8: 177–195.

Ralls, K. 1967. Auditory sensitivity in mice, *Peromyscus* and *Mus musculus*. Animal Behaviour 15: 123–128.

————. 1971. Mammalian scent marking. Science (New York) 171: 443–449.

Randolph, J. C. 1973. Ecological energetics of a homeothermic predator, the short-tailed shrew. Ecology 54: 1166–1187.

Rasa, O. E. A. 1973a. Marking behavior and its social significance in the African dwarf mongoose, *Helogale undulata rufula*. Zeitschrift für Tierpsychologie 293–318.

————. 1973b. Prey capture, feeding techniques, and their ontogeny in the African dwarf mongoose. Zeitschrift für Tierpsychologie 32: 449–488.

Rohwer, S. A. and D. L. Kilgore, Jr. 1973. Inbreeding in the arid-land foxes, *Vulpes velox* and *V. macrotis*. Systematic Zoology 22: 157–165.

Roitblat, H. L., T. G. Bever, and H. S. Terrace. 1983. Animal Cognition. Erlbaum Inc. Hillsdale, New Jersey.

Romer, A. S. 1962. *The Vertebrate Body*. 3rd. ed. W. B. Saunders Co. Philadelphia.

————. 1966. *Vertebrate Paleontology*. 3rd. ed. University of Chicago Press. Chicago.

Rowe, J. S. 1959. Forest regions of Canada. Canada Dept. of Northern Affairs and Natural Resources, Forestry Branch, Bulletin No. 123. Ottawa.

Rowley, I. 1963. "Rodent run" distraction by a passerine, the Superb Wren. Behaviour 20: 170–176.

Rue, L. L. III. 1969. *The World of the Red Fox*. J. B. Lippincott. Philadelphia.

Sande, J. O. 1943. Kvordan reven hamstrer om vinteren. Naturen: 255–256.

Sargeant, A. B. 1972. Red fox spatial characteristics in relation to waterfowl predation. Journal of Wildlife Management. 36: 225–248.

————. 1978. Red fox prey demands and implications to prairie duck production. Journal of Wildlife Management. 42: 520–527.

———— and L. E. Eberhardt. 1975. Death feigning by ducks in response to predation by red foxes *(Vulpes fulva)*. Amer. Midl. Naturalist: 94: 108–119.

Savage, A. and C. Savage. 1981. *Wild Mammals of Western Canada*. Western Producer Prairie Books. Saskatoon, Canada.

Savage, R. J. G. 1977. Evolution in carnivorous mammals. Palaeontology 20: 237–271.

Scapino, R. 1981. Morphological investigation into functions of the jaw symphysis in Carnivorans. Journal of Morphology 67: 339–375.

Schaller, G. B. 1967. *The Deer and The Tiger*. University of Chicago Press. Chicago.

————. 1972. *The Serengeti Lion*. University of Chicago Press. Chicago.

Schantz, T. von. 1980. Prey consumption of a red fox population in southern Sweden. Biogeographica 18: 53–64.

————. 1981. Female cooperation, male competition, and dispersal in the red fox, *Vulpes vulpes*. Oikos 37: 63–68.

Schauenberg, P. 1970. Le chat forestier d'Europe *Felis s. silvestris* Schreber 1797 en Suisse. Revue Suisse de Zoologie 77: 127–190.

Schenkel, R. 1966. Zum Problem der Territorialität und des Markierens bei Säugern—am Beispiel des Schwarzen Nashorns und des Löwens. Zeitschrift für Tierpsychologie 23: 593–626.

Schofield, R. D. 1960. A thousand miles of fox trails in Michigan's ruffed grouse range. Journal of Wildlife Management 24: 432–434.

Schueler, R. L. 1951. Red fox food habits in a wilderness area. Journal of Mammalogy 32: 462–464.

Schultze-Westrum, T. G. 1965. Innerartliche Verständigung durch Düfte beim Gleitbeutler *Petaurus breviceps papuanus* Thomas (Marsupialia, Phalangeridae). Zeitschrift für Vergleichende Physiologie. 38: 84–135.

———. 1969. Social communication by chemical signals in flying phalangers. *In:* C. Pfaffmann (ed.) *Olfaction and Taste. III* Rockefeller University Press. New York.

Scott, T. G. 1943. Some food coactions of the northern plains red fox. Ecological Monographs 13: 427–479.

———. 1947. Comparative analysis of red fox feeding trends on two central Iowa areas. Iowa Agricultural Experimental Station, Resource Bulletin No. 353: 426–487.

——— and W. D. Klimstra. 1955. Red foxes and a declining prey population. South Illinois University Monograph Series No. 1: 1–123.

Seagears, C. B. 1944. The red fox in New York. New York Conservation Department Education Bulletin 83p.

Sebeok, T. A. and R. Rosenthal. 1981. The Clever Hans phenomenon: Communication with horses, apes, and people. Annals of the New York Academy of Science 364: 1–311.

——— and J. Umiker-Sebeok. (eds.) 1980. *Speaking of Apes: A Critical Anthology of Two-way Communication with Man.* Plenum Press. New York.

Seidensticker, J. C. IV, M. G. Hornocker, W. V. Wiles, and J. P. Messick. 1973. Mountain Lion social organization in the Idaho Primitive area. Wildlife Monographs 35: 1–60.

Seitz, A. 1950. Untersuchungen über angeborene Verhaltensweisen bei Caniden. I und II. Zeitschrift für Tierpsychologie 7: 1–46.

Seitz, E. 1969. Die Bedeutung geruchlicher Orientierung beim Plumplori, *Nycticebus coucang* Boddaert 1785. (Prossimii, Lorisidae). Zeitschrift für Tierpsychologie 26: 73–103.

Seton, E. T. 1929. *Lives of Game Animals.* Doubleday and Co. Inc. New York.

Sheldon, W. G. 1950. Denning habits and home range of red foxes in New York State. Journal of Wildlife Management 14: 33–42.

Simmons, K. E. L. 1952. The nature of the predator reactions of breeding birds. Behaviour 4: 161–172.

———. 1955. The nature of the predator reactions of waders towards humans with special reference to the role of the aggressive-escape-and brooding drives. Behaviour 8: 130–173.

Simpson, G. G. 1945. The principles of classification and a classification of mammals. Bulletin of the American Museum of Natural History 85: 1–350.

Southern, H. N. and J. S. Watson. 1941. Summer food of the red fox *(Vulpes vulpes)* in Great Britain: a preliminary report. Journal of Animal Ecology 10: 1–11.

Speller, S. W. 1972. Food ecology and hunting behaviour of denning Arctic foxes at Aberdeen Lake, Northwest Territories. Ph.D. Thesis. University of Saskatchewan.

Spiegel, M. R. 1967. *Theoretical Mechanics.* McGraw-Hill Book Co. New York.

Sprague, R. H. and J. J. Anisko. 1973. Elimination patterns in the laboratory beagle. Behaviour 47: 257–267.

Stager, K. E. 1964. The role of olfaction in food location by the turkey vulture *(Cathartes aura).* Los Angeles County Museum Contribution in Science 81: 1–63.

———. 1967. Avian olfaction. American Zoologist 7: 415–419.

Stains, H. J. 1967. Carnivores and pinnipeds. *In:* S. Anderson and J. Knox Jones, Jr., (eds.) *Recent Mammals of the World.* Ronald Press. New York.

———. 1975. Distribution and taxonomy in Canidae. *In:* M. W. Fox (ed.) *The Wild Canids.* Van Nostrand Reinhold Co. New York.

Steck, F., A. Wandeler, P. Bichsel, S. Capt, and L. Schneider. 1982. Oral immunisation of foxes against rabies: A field study. Zentralblatt Veterinärmedizin. Reihe B. 29: 372–396.

Storm, G. L. 1965. Movements and activities of foxes as determined by radiotracking. Journal of Wildlife Management. 29: 1–13.

———. 1972. Population dynamics of red foxes in northcentral United States. Ph.D. Thesis. University of Minnesota.

———. and G. G. Montgomery. 1975. Dispersal and social contact among red foxes: results from telemetry and computer simulation. *In:* M. W. Fox (ed.) *The Wild Canids.* Van Nostrand Reinhold Co. New York.

Storm, G. L., R. D. Andrews, R. L. Phillips, R. A. Bishop, D. B. Siniff, and J. R. Tester. 1976. Morphology, reproduction, dispersal, and mortality of midwestern red fox populations. Wildlife Monographs 49: 1–82.

Stubbe, M. 1980. Population ecology of the red fox—*Vulpes vulpes* (L. 1758)—in the G. D. R. Biogeographica 18: 71–96.

Swan, M. and R. L. Dix. 1966. The phytosociological structure of upland forest at Candle Lake, Saskatchewan. Journal of Ecology 54: 13–40.

Swink, F. N. 1952. Effects of the red fox on other game species. Virginia Wildlife 13: 20–22.

Taketazu, M. 1979. *Fox Family: Four Seasons of Animal Life.* John Weatherhill Inc. of Tokyo.

Taylor, A. 1935. Skeletal changes associated with increasing body size. Journal of Morphology 57: 253–274.

Tembrock, G. 1957a. Zur ethologie des Rotfuchses (*Vulpes vulpes* L) unter besonderer Berucksichtigung der Fortpflanzung. Der Zoologische Garten 23: 289–532.

———. 1957b. Das Verhalten des Rotfuchses. Handbook of Zoology. VIII Part 10 15: 1–20.

———. 1958a. Aktivatsperiodik bei *Vulpes* and *Alopex.* Zoologische Johrbuecher 68: 297–324.

———. 1958b. Spielverhalten beim Rotfuchses. Zodogische Beitraege 3: 423–496.

———. 1959. Beobachtungen zur Fuchspranz unter besonderer Berücksichtigung der Lautgeburg. Zeitschrift für Tierpsychologie 16: 351–368.

———. 1963. Muschlaute beim Rotfuchs (*Vulpes vulpes* L.). Zeitschrift für Tierpsychologie 20: 616–623.

Thompson, D. W. 1942. *On Growth and Form.* 2nd ed. Cambridge University Press. Cambridge, England.

Thornton, W. A., G. C. Creel, and R. E. Trimble. 1971. Hybridization in the fox genus *Vulpes* in west Texas. Southwest Naturalist 15: 473–484.

——— and G. C. Creel. 1975. The taxonomic status of kit foxes. Texas Journal of Science 26: 127–136.

Tinbergen, N. 1951. *The Study of Instinct.* Clarendon Press. Oxford, England.

———. 1962. Foot paddling in gulls. British Birds 55: 117–120.

———. 1963a. On aims and methods of ethology. Zeitschrift für Tierpsychologie 20: 410–433.

———. 1963b. Behavior and natural selection. *In:* J. A. Moore (ed.) *Ideas in Modern Biology.* Harper and Row. New York.

———. 1965. Von der Vorralskammern des Rotfuchses (*Vulpes vulpes* L.). Zeitschrift für Tierpsychologie 22: 119–149.

————. 1966. Adaptive features of the black-headed gull, *Larus ridibundus* L. Proceedings of the XIV International Ornithological Congress: 43–59.

————. 1969a. Introduction to the 1969 reprint. *The Study of Instinct.* Clarendon Press. Oxford. England.

————. 1969b. On war and peace in animals and man. Science (New York) 170: 1411–1418.

————, G. J. Broekhuysen, F. Feekes. J. C. W. Houghton, and E. Szulc. 1962a. Egg shell removal by the black-headed gull, *Larus ridibundus* L.: a behavioural component of camouflage. Behaviour 19: 74–117.

————, H. Kruuk, M. Pailette, and R. Stamm. 1962b. How do black-headed gulls distinguish between eggs and egg-shells? British Birds 55: 120–129.

————, M. Impekoven, and D. Franck. 1967. An experiment on spacing out as a defense against predation. Behaviour 28: 307–321.

Toldt, K. 1907. Studien über das Haarkleid von *Vulpes vulpes* L. Annalen für Naturhistorie Hofmuseum 22: 197–269.

Trapp, G. and D. L. Hallberg. 1975. Ecology of the gray fox *(Urocyon cinereoargenteus):* A Review. *In:* M. W. Fox (ed.) *The Wild Canids.* Van Nostrand Reinhold Co. New York.

Treisman, M. 1975a. Predation and the evolution of gregariousness. I. Models for concealment and evasion. Animal Behaviour 23: 779–800.

————. 1975b. Predation and the evolution of gregariousness. II. An economic model for predator-prey interaction. Animal Behaviour 23: 801–825.

Trivers, R. L. 1971. Evolution of reciprocal altruism. Quarterly Review of Biology 46: 35–57.

————. 1974. Parent-offspring conflict. American Zoologist 14: 249–264.

Troughton, E. 1957. *Furred Animals of Australia.* Government Printer. Adelaide, Australia.

Van Tyne, J. and A. J. Berger. 1959. *Fundamentals of Ornithology.* John Wiley and Sons. New York.

Vesey-Fitzgerald, B. 1966. *Town Fox, Country Fox.* Andre Deutsch Ltd. London.

Vincent, L. E. and M. Bekoff. 1978. Quantitative analysis of the ontogeny of predatory behavior in coyotes, *Canis latrans.* Animal Behaviour 26: 225–231.

Voigt, D. R. and B. D. Earle. 1983. Avoidance of coyotes by red fox families. Journal of Wildlife Management 47: 852–857.

Walker, E. P. 1978. *Mammals of the World.* Johns Hopkins Press. Baltimore.

Walker, S. 1983. *Animal Thought.* Routledge and Kegan Paul Ltd. London.

Walls, G. L. 1967. *The Vertebrate Eye and Its Adaptive Radiation.* Hafner. New York.

Walther, F. R. 1969. Flight behaviour and avoidance of predators in Thomson's gazelles. Behaviour 34: 184–221.

Wenzel, B. M. 1967. Olfactory perception in birds. *In:* T. Kayashi (ed.) *Olfaction and Taste II.* Pergamon Press. London.

Whitten, W. K. and F. H. Bronson. 1970. The role of pheromones in mammalian reproduction. *In:* J. W. Johnston, Jr., D. G. Moulton, and A. Turk (eds.) *Advances in Chemoreception I. Communication by Chemical Signals.* Appleton, Century, Crofts, Co. New York.

Wickler, W. 1967. Vergleichende Verhaltensforschung und Phylogenetik. *In:* G. Heberer (ed.) *Die Evolution der Organismen I.* 3rd ed. G. Fischer. Stuttgart.

————. 1968. *Mimicry in Plants and Animals.* Weidenfeld and Nicolson. London.

Wilson, E. O. 1975. *Sociobiology: The New Synthesis.* Harvard University Press. Cambridge. Massachusetts.

Winton, W. E. de. 1899. Exhibition of, and remarks upon, the tail of a fox *(Canis vulpes)* showing the gland on the upper surface. Proceedings of the Zoological Society (London) 1899: 292–295.

Wray, G. 1976. The functional anatomy of the digital anatomy in *Vulpes vulpes fulva* and comparisons with other representative Carnivorans. Honors Thesis (Unpublished) University of Calgary. 87 pp.

Wurster, D. H. and K. Benirschke. 1968. Comparative studies in the order Carnivora. Chromosoma 24: 336–382.

Yoneda, M. 1982. Effects of hunting on age structure and survival rates of red fox in eastern Hokkaido. Journal of Wildlife Management 46: 781–786.

Young, S. P. and H. H. T. Jackson. 1951. *The Clever Coyote*. Parts I and II. Stackpole Co. Harrisburg, Pennsylvania.

Zimen, E. (ed.) 1980a. The red fox: A symposium on behaviour and ecology. Biogeographica 18: 1–285. Dr. W. Junk b.v. Publishers. The Hague.

———. 1980b. Introduction: A short history of human attitudes towards the fox. Biogeographica 18: 1–6.

———. 1980c. Rabies control. Biogeographica 18: 277–285.

———. 1984. Long range movements of the red fox, *Vulpes vulpes* L. Acta Zoologica Fennica No. 171: 267–270.

Index

Photo credits

Color: Joe Benge, Parks Canada, An intelligent, small predator; R. L. Dutcher, Parks Canada, Between four and six weeks of age; Brian Milne/First Light, By six months of age.

Black and white: Mervyn Syroteuk, Parks Canada, p. 29; Stephen J. Krasemann/DRK Photo, p. 41; Esther Schmidt, p. 53; R. L. Dutcher, Parks Canada, p. 62; Dennis W. Schmidt, pp. 65, 67, 79; Dennis Dyck, pp. 68, 76; Joe Benge, Parks Canada, p. 101.

All sketches by Mary Romanuck, Prairie Wildlife Art, R.R. No. 6, Saskatoon, Canada.

All other photographs were taken by the author.